PROTEINS AND ENZYMES
Applications

MJP
PUBLISHERS

PROTEINS AND ENZYMES
Applications

R. PUVANAKRISHNAN

Emeritus Scientist, Former Director Grade Scientist and Head
Department of Biotechnology, CSIR-Central Leather Research Institute
Adyar, Chennai - 600 020.

T. HEMALATHA

Scientist, DST-Women Scientist Programme
CSIR-Central Leather Research Institute
Adyar, Chennai - 600 020.

T. UMAMAHESWARI

Department of Food Science and Nutrition
Community Science College and Research Institute, TNAU
Madurai - 625 104.

Chennai　　　Trichy　　　Tirunelveli　　　New Delhi

Cover image description : Structure of Hemoglobin

PREFACE

Proteins are the most versatile and abundant macromolecules present in all living cells. They are linear polymers built of monomer units called amino acids. They contain a wide range of functional groups and can interact with one another and with other biological molecules to form complex assemblies. Proteins are formed by condensing the α-amino group of one amino acid (imino group in the case of proline) with the α-carboxyl group of another, with the concomitant loss of a water molecule and the formation of a peptide bond. They function as catalysts, provide immune protection, transmit nerve impulses, control growth and differentiation etc.

Enzymes are biological catalysts characterized by their extraordinary specificity and reactivity in living systems. They are found in all living organisms where they catalyse and regulate innumerable chemical reactions. Further, researchers on the biochemistry of living organisms have demonstrated the importance of enzymes and enzymology has become one of the essential sciences of the present day. Enzymes have immense applications both in industry and medicine. The global market for medical enzymes has been estimated at US $ 6.0 billion in 2010 and it is expected to grow at an annual rate of about 4.0 % to reach around US $ 8.5 billion in 2020.

In 2015, the authors brought about a monograph entitled "Microbes and Enzymes – Basics and Applied" wherein a chapter was dedicated for the medical applications of enzymes. This became the basic thought for conceiving the present book. Many reviews have been published on the applications of proteins and enzymes in medicine and industry. But, the information available is scattered and there is an urgent need for a cohesive presentation of the role of proteins and enzymes in diagnosis and therapy.

For conceiving this project, the medical applications of proteins and enzymes are categorised as follows: (i) proteins in diagnosis, (ii) proteins in therapy, (iii) enzymes in diagnosis and (iv) enzymes in therapy. This book is meant for students taking up paramedical courses having biochemistry as one of the components. Specifically, the target audience will be the undergraduate students of biochemistry, medical laboratory technology, nutrition, biotechnology, microbiology, nursing and pharmacy.

This book will have seven chapters. The properties and characteristics of proteins are to be properly understood for developing them as suitable biomarkers and drugs. Hence, the basics of amino acids, structure, classification, denaturation and renaturation of proteins and assay procedures for protein estimation are dealt with in Chapter 1. To develop enzymes optimally for use in diagnosis and therapy, a proper understanding of their properties is absolutely necessary. Therefore, Chapter 2 details the classification, specificity, factors affecting enzyme activity, isoenzymes, Michaelis-Menten Kinetics and enzyme inhibition. Chapter 3 dwells on the use of protein markers in different areas viz., cardiology, diabetology, hematology, oncology, hepatology etc. Chapter 4 talks about how therapeutic proteins are useful as drugs in endocrine disorders, metabolic enzyme deficiencies, pulmonary and gastrointestinal tract disorders, immunoregulation etc. Chapter 5 underlines how enzymes are useful as biomarkers for diagnosing different diseases. Chapter 6 focuses on the use of enzymes as therapeutic agents in various diseases viz. cancer, pancreatic disorders, heart diseases, kidney diseases etc. Chapter 7 tries to portray the new proteins and enzymes which could be developed as future markers and drugs. A Glossary is added wherein all the important terms used in this book are defined and this will help the student to understand the subject matter better. A list of abbreviations is also provided along with an index.

Our heart felt thanks go to all those who helped us in this stupendous task. Profuse thanks go to **Dr. B.R. Nammalwar**, Senior Pediatric Nephrologist, Chennai for his kind encouragement, critical suggestions and specifically his inputs on "Protein markers". Our heart felt thanks go to **Prof. G. Selvaraj**, Formerly of the Department of Biochemistry, Madras Medical College, Chennai, **Dr. M. Balasubramaniam**, Dean of Research

Studies, Madras Diabetes Research Foundation, Chennai, **Dr. P.S. Jagdish**, Consultant, Dr. Mohan's Diabetes Specialities Centre, Chennai, **Dr. Mallika Ravindran,** Director - Lab, Anderson Diagnostic Labs, Chennai and **Mr. Amritraj**, Biochemist, V.S. Hospitals, Chennai, for their kind inputs on "Proteins and Enzyme Markers".

We thank our friends from the academia who are kind enough to review the book chapters.

Dr. G. Jayamathi	**Dr. S. Sivakumar**	**Dr. N.S. Kaviyarasi**
Professor	Department of Biochemistry	Department of Chemistry
Department of Biochemistry	Sri Sankara College of Arts and Science	Mount Carmel College, Bengaluru – 560 052
Sri Balaji Dental College, Chennai – 600 100	Enathur, Kanchipuram – 631 561	

The authors thank profusely their family members for their kind cooperation and support. Any suggestions or positive criticism are welcome and may be sent to puvanakrishnan@yahoo.com.

R. Puvanakrishnan

T. Hemalatha

T. UmaMaheswari

ABBREVIATIONS

A1AT	Alpha 1 antitrypsin
ACP	Acid phosphatase
ACPA	Anti-cyclic citrullinated peptide antibodies
AFP	Alpha feto protein
AIDS	Acquired Immuno Deficiency Syndrome
ALP	Alkaline phosphatase
ALT	Alanine transaminase
ANA	Anti-nuclear antibodies
APA	Antiphospholipid antibodies
apo A	Apolipoprotein A
apo B	Apolipoprotein B
ASA	Anti-Sperm antibodies
ASO	Antistreptolysin O
AST	Aspartate aminotransferase
ATP	Adenosine triphosphate
BMP	Bone morphogenic protein
BNP	B-type natriuretic peptide
BSA	Bovine Serum Albumin

CA 125	Cancer antigen 125
CBER	Center for Biologics Evaluation and Review
CCs	Cysteine cathepsins
CD	Cluster of differentiation
CDER	Center for Drug Evaluation and Review
CEA	Carcinoembryonic antigen
CK	Creatine kinase
COX	Cyclooxygenase
Cp	Ceruloplasmin
CRP	C – reactive protein
cTnI	Cardiac troponins I
cTnT	Cardiac troponins T
DM	Diabetes mellitus
DNA	Deoxyribonucleic acid
dsDNA/RNA	Double stranded DNA/RNA
EC	Enzyme Commission
ECL	Enhanced Chemi Luminescence
ECM	Extracellular matrix
EF	Elongation factor
EGFR	Epidermal growth factor receptor
Epo	Erythropoietin
ERT	Enzyme replacement therapy
ES complex	Enzyme-Substrate Complex
FAD	Flavin adenine dinucleotide

FSH	Follicle stimulating hormone
G6PD	Glucose-6-phosphate dehydrogenase
GAG	Glycosaminoglycans
GC	Gas chromatography
GCSF	Granulocyte colony stimulating factor
GGT	Gamma glutamyl transferase
GH	Growth hormone
GHRH	Growth hormone releasing hormone
GLP1	Glucagon like peptide 1
GnRH	Gonadotropin releasing hormone
GPBB	Glycogen phosphorylase BB
GPH	Glycoprotein hormone
HAV	Hepatitis A virus
Hb	Hemoglobin
HbA1c	Glycosylated hemoglobin
HBV	Hepatitis B virus
hCG	Human chorionic gonadotropin
HCl	Hydrochloric acid
HCV	Hepatitis C virus
HDL	High density lipoproteins
HDN	Haemolytic disease of the newborn
HIV	Human immune deficiency virus
HPLC	High Performance Liquid Chromatography
IFN	Interferon

Ig	Immunoglobulin
IGF-1	Insulin like growth factor - 1
IL-2	Interleukin-2
KGF	Keratinocyte growth factor
LDH	Lactate dehydrogenase
LDL	Low density lipoproteins
LH	Luteinizing hormone
LPS	Lipopolysaccharide
LSDs	Lysosomal storage diseases
Mabs	Monoclonal antibodies
MI	Myocardial infarction
MMP	Matrix metalloproteinase
NAD	Nicotinamide Adenine Dinucleotide
NADP	Nicotinamide Adenine Dinucleotide Phosphate
NHL	Non-Hodgkin's lymphomas
NSAID	Nonsteroidal anti-inflammatory drugs
NT pro BNP	N – terminal pro B-type natriuretic peptide
PDGF	Platelet-derived growth factor
PEG	Polyethylene glycol
PERT	Pancreatic enzyme replacement therapy
PG	Proteoglycans
PSA	Prostate specific antigen
PTH	Parathyroid hormone
RA	Rheumatoid arthritis

RBCs	Red blood corpuscles
RF	Rheumatoid factor
Rh	Rhesus
RNA	Ribonucleic acid
RSV	Respiratory syncytial virus
SAA	Serum amyloid A
SDH	Succinate dehydrogenase
SDS	Sodium Dodecyl Sulphate
SDS-PAGE	Sodium dodecyl sulphate-polyacrylamide gel electrophoresis
SGPT	Serum glutamate pyruvate transaminase
SGOT	Serum glutamate oxaloacetate transaminase
SHBG	Sex hormone binding globulin
SK	Streptokinase
SLE	Systemic lupus erythematosus
TB	Tuberculosis
TM	Trade Mark
TNF-α	Tumor necrosis factor - α
TPA	Tissue-plasminogen activator
TRAP	Tartrate-resistant acid phosphatase
TSH	Thyroid stimulating hormone
uPA	urokinase-type plasminogen activator
USFDA	United States Food and Drug Administration
UV	Ultraviolet

VEGF	Vascular endothelial growth factor
VLDL	Very low density lipoproteins
vWF	von Willebrand factor
WBCs	White blood corpuscles
β2M	Beta 2 microglobulin

CONTENTS

CHAPTER 1

PROTEINS - BASICS

INTRODUCTION

Proteins are the most versatile and abundant macromolecules present in all living cells. They are formed by condensing the α-amino group of one amino acid or the imino group of proline with the α-carboxyl group of another, with the concomitant loss of a molecule of water and the formation of a peptide bond. They function as catalysts, provide immune protection, transmit nerve impulses, control growth and differentiation etc.

$$^+H_3N-\underset{R_1}{\overset{H}{C}}-C\diagdown^O_{O^-} \; + \; ^+H_3N-\underset{R_2}{\overset{H}{C}}-C\diagdown^O_{O^-} \longrightarrow \; ^+H_3N-\underset{R_1}{\overset{H}{C}}-\overset{O}{C}-\underset{R_2}{\overset{H}{N}-\overset{}{C}}-C\diagdown^O_{O^-} \; + \; H_2O$$

Peptide bond

Proteins are linear polymers built of monomer of units called aminoacids. It contains a wide range of functional groups and can interact with one another and with other biological molecules to form complex assemblies. The progressive condensation of many molecules of amino acids gives rise to an unbranched polypeptide chain. By convention, the N-terminal amino acid is taken as the beginning of the chain with the C-terminal amino acid as the end of the chain (proteins are biosynthesized in this direction). Polypeptide chains contain between 20 and 2000 amino acid residues and hence, have a relative molecular mass ranging between 2000 and 2,00,000. Many proteins have a relative molecular mass in the range 20000 to 100000. The distinction between a large peptide and a small protein is not clear. Generally, chains of amino acids containing fewer than 50 residues are referred to as peptides and those with more than 50 are referred to as proteins. Most proteins contain many hundreds of amino acids (ribonuclease is an extremely small protein with only 103 amino acid residues) and many biologically active peptides contain 20 or fewer amino acids, e.g. oxytocin (9 amino acid residues), vasopressin (9), enkephalins (5), gastrin (17), somatostatin (14) and luteinising hormone (10).

AMINO ACIDS

Amino acids are the basic building blocks of proteins and they play an important role as intermediates in metabolism. There are 20 different amino acids found in living cell. The first to be discovered was asparagine, in 1806. The last of the 20 to be found, threonine, was not identified until 1938. From these building blocks, different organisms construct widely diverse products as enzymes, hormones, antibodies, transporters, muscle fibers, feathers, milk proteins, antibiotics, mushroom and innumerable other substances having distinct biological activities. The chemical properties of the amino acids of a protein determine the biological activity of the protein. All biological amino acids have the basic skeleton

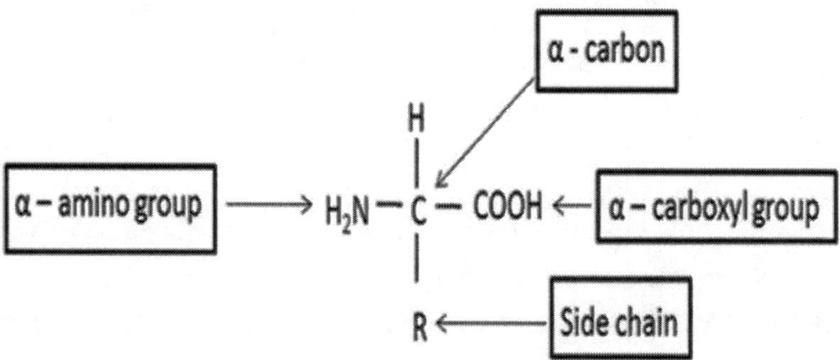

All amino acids, except glycine, have a central chiral carbon atom with 4 attachments viz., a single hydrogen atom, a COOH (α - carboxyl) group, a NH_2 (α - amino) group and a side chain (R) which differs for each of the

20 amino acids. Amino acid, derived from the presence of the amino group and carboxyl group, dissociates in water to give COO^- and H^+ ions and it is acidic because of the presence of H^+ ions in the solution. The 20 amino acids are defined by the nature of the side chain R. The chemical properties of the amino acids of a protein determine the biological activity of the protein. The 20 different types of amino acids are shown in Table 1.1 and Figure 1.1.

Table 1.1   Different types of Amino Acids

Amino acid	Three letter symbol	One letter symbol
Alanine	Ala	A
Arginine	Arg	R
Asparagine	Asn	N
Aspartic acid	Asp	D
Cysteine	Cys	C
Glutamic acid	Glu	E
Glutamine	Gln	Q
Glycine	Gly	G
Histidine	His	H
Isoleucine	Ile	I
Leucine	Leu	L
Lysine	Lys	K
Methionine	Met	M
Phenylalanine	Phe	F
Proline	Pro	P
Serine	Ser	S
Threonine	Thr	T

Amino acid	Three letter symbol	One letter symbol
Tryptophan	Trp	W
Tyrosine	Tyr	Y
Valine	Val	V

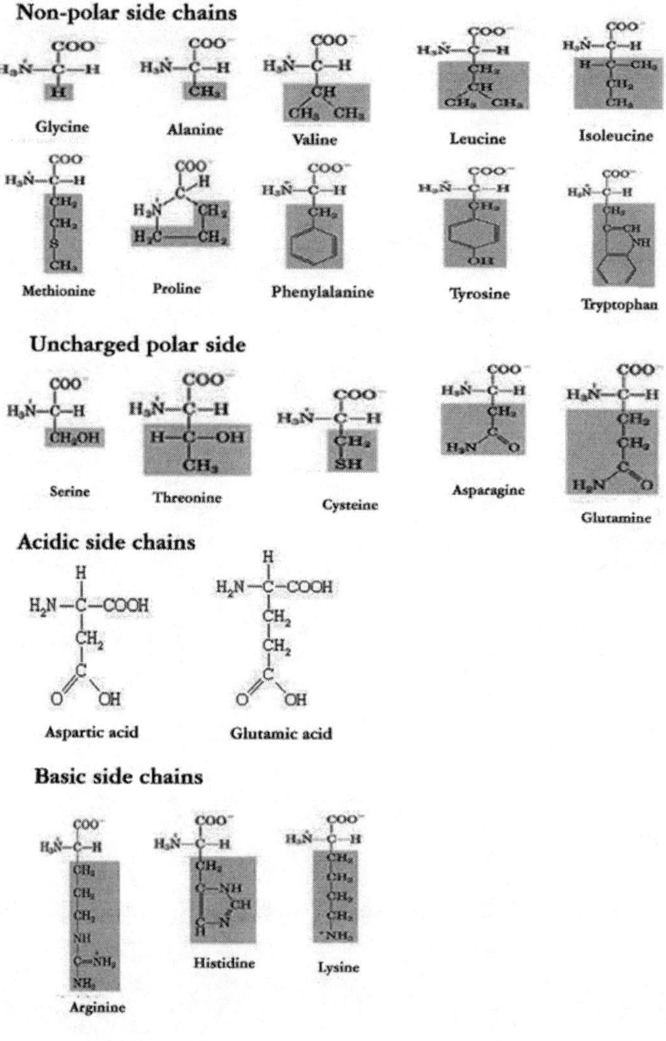

Figure 1.1   Structure of amino acids

Among the 20 different amino acids, glycine is the simple amino acid with a single hydrogen atom as its side group. Tyrosine, tryptophan and phenylalanine are aromatic amino acids, i.e. they contain a aromatic ring as a part of their side chain, while methionine and cysteine contain sulfur in their side chain. The amino acid residues in protein molecules are exclusively L-stereoisomers. D-amino acid residues have been found only in a few, generally small peptides, including some peptides of bacterial cell walls and certain peptide antibiotics. Cells are able to specifically synthesize the L-isomers of amino acids because the active sites of enzymes are asymmetric; causing the reactions they catalyze to be stereospecific. When amino acids are dissolved in water at neutral pH, they act as either an acid or a base. This property is referred to as zwitterion or dipolar ion. Amino acids vary in their acid-base properties and have characteristic titration curves and isoelectric point/pH.

The amino acids are grouped into four types based on their polarity i.e., their tendency to interact with water at biological pH. The polarity of R groups may vary from nonpolar and hydrophobic to polar and hydrophilic. The four types include:

i)   (a) Nonpolar aliphatic and hydrophobic R groups include glycine, alanine, valine, leucine, isoleucine, methionine and proline.

   (b) Aromatic R groups are relatively nonpolar and hydrophobic. It includes phenylalanine, tyrosine and tryptophan.

ii)   Polar uncharged R groups are more hydrophilic than nonpolar groups. These aminoacids are serine, threonine, cysteine, asparagine and glutamine.

iii)   Polar negatively charged R groups include acidic amino acids such as aspartic acid and glutamic acid.

iv)   Polar charged R groups are positively charged, hydrophilic basic group amino acids. They are lysine, arginine and histidine.

In addition to the 20 common amino acids, a few uncommon amino acids are present due to the modification of common residues. They are 4-hydroxyproline, a derivative of proline, 5-hydroxylysine, derived from lysine, 6-*N*-methyllysine, a constituent of myosin, γ-carboxyglutamic acid, found in prothrombin, desmosine, a derivative of four lysine residues and selenocysteine, derived from serine. Some 300 additional amino acids have been found in cells. They have a variety of functions but are not constituents of proteins.

Essential and Nonessential Amino Acids

Humans can produce 10 of the 20 amino acids. The others must be supplied in the food. Failure to obtain enough of even 1 of the 10 essential amino acids, those that we cannot make, results in degradation of the body's proteins. Unlike fat and starch, the human body does not store excess amino acids for later use; the amino acids must be in the food every day.

Nonessential amino acids: The 10 amino acids that are synthesized in the living system are asparagine, aspartic acid, cysteine, glutamic acid, glutamine, glycine, proline, serine and tyrosine. Tyrosine is produced from phenylalanine; so, if the diet is deficient in phenylalanine, tyrosine will be required as well. Arginine is required in the diet for the young, but not for adults.

Essential amino acids: They are not synthesized in the human system and have to be supplemented in the diet. A simple formula viz. MATTVILPHLY helps to know the essential amino acids and they are: **M**ethionine, **A**lanine, **T**hreonine, **T**ryptophan, **V**aline, **I**soleucine, **L**eucine, **P**henylalanine, **H**istidine and **L**ysine.

PEPTIDES AND PROTEINS

Two amino acid molecules can be covalently coupled through a substituted amide linkage (peptide bond) to yield a dipeptide. Peptide bond is formed by removal of the elements of water from the α-carboxyl group of one amino acid and the α-amino group of another. Three amino acids can be joined by two peptide bonds to form a tripeptide; similarly, amino acids can be linked to form tetrapeptides, pentapeptides, and so on. A few amino acids are joined to form oligopeptide. When many amino acids are joined, the structure is called polypeptide, which is otherwise called as protein. Naturally occurring peptides range in length from two to many thousands of amino acid residues. Like free amino acids, peptides have characteristic titration curves and a characteristic isoelectric pH at which they do not move in an electric field. This property can be used to separate protein and this method is known as isoelectric focusing.

PROTEIN STRUCTURE

The spatial arrangement of atoms in a protein is called its conformation. Any conformation of a protein can be achieved without breaking covalent bonds. Proteins in any of their functional, folded conformations are called native proteins. Each protein possesses a characteristic three dimensional shape, its conformation. There are four separate levels of structure and organization (Figure 1.2) as follows:

i) Primary, ii) Secondary, iii) Tertiary and iv) Quaternary

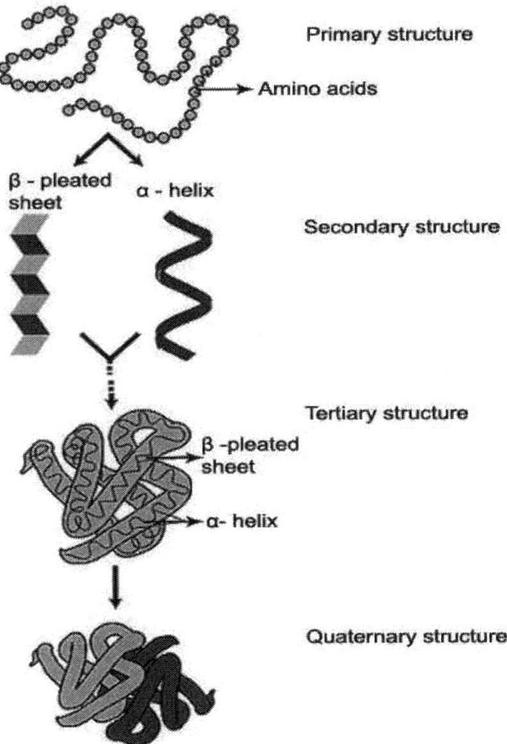

Figure 1. 2   STRUCTURE OF PROTEIN

Primary Structure

The primary structure of a protein defines the sequence of the amino acid residues and it is dictated by the base sequence of the corresponding gene(s). Indirectly, the primary structure also defines the amino acid composition (which of the possible 20 amino acids are actually present) and content (the relative proportions of the amino acids present).

Secondary Structure

Secondary structure defines the localized folding of some part of the polypeptide chain due to hydrogen bonding. Polypeptide chains can be folded into regular patterns such as the α-helix and β-pleated sheet. Some uncommon folding structures such as β-turns and Ω-loop are also identified. The α-helix is a coiled structure stabilized by intra-chain hydrogen bonds whereas β-sheets are stabilized by hydrogen bonding between antiparallel polypeptide strands. The polypeptide chains can change direction by making reverse turns (also known as the β-turn or hairpin bend) and loops (omega loop). Turns and loops invariably lie on the surfaces of proteins and thus often participate in interactions between proteins and other molecules. The distribution of α-helices, β-strands, and turns along a protein chain is often referred to as its secondary structure. A typical feature of α-helical structure is that it contains more number of nonpolar amino acid residues. Membrane bound proteins possess α-helical structure and they are engaged in hydrophobic interactions with cell membrane. Some of the 20 amino acids found in proteins, including proline, isoleucine, tryptophan and asparagine, disrupt α-helical structures. Some proteins have up to 70% secondary structure but others have none.

Tertiary Structure

Tertiary structure defines the overall folding of a polypeptide chain i.e. water-soluble proteins fold into compact structures with nonpolar cores. It is stabilized by electrostatic attractions between oppositely charged ionic groups ($-NH_3^+$, COO^-), weak van der Waals forces, hydrogen bonding, hydrophobic interactions and, in some proteins, by disulphide (-S — S-)

bridges formed by the oxidation of spatially adjacent sulphydryl groups (-SH) of cysteine residues. The three-dimensional folding of polypeptide chains is such that the interior consists predominantly of non-polar, hydrophobic amino aid residues such as valine, leucine and phenylalanine. The polar, ionized, hydrophilic residues are found on the outside of the molecule, where they are compatible with the aqueous environment. However, some proteins also have hydrophobic residues on their outside and the presence of these residues is important in the processes of ammonium sulphate fractionation and hydrophobic interaction chromatography.

Quaternary Structure

Polypeptide chains can be assembled into multi subunit structures and the spatial arrangement of subunits and the nature of their interactions are referred to as Quaternary structure. It is restricted to oligomeric proteins, which consist of the association of two or more polypeptide chains held together by electrostatic attractions, hydrogen bonding, van der Waals forces and occasionally disulphide bridges. Thus, disulphide bridges may exist within a given polypeptide chain (intra-chain) or linking different chains (inter-chain). An individual polypeptide chain in an oligomeric protein is referred to as a subunit. The subunits in a protein may be identical or different: e.g. haemoglobin consists of two α- and two β-chains, and lactate dehydrogenase of four (virtually) identical chains.

PROTEIN CLASSIFICATION

Because of the complexity of protein molecules and their diversity of function, it is very difficult to classify them in a single, well-defined fashion. Three alternative methods of classification are:

i.   According to structure – globular, fibrous and intermediate proteins

ii. According to composition – simple and conjugated (Table 1.2)

iii. According to function – structural proteins, enzymes, antibodies, hormones etc. (Table 1.3)

Globular, Fibrous and Intermediate Proteins

Globular Proteins are approximately spherical in shape, are generally water soluble and may contain a mixture of α-helix, β-pleated sheet and random structures. Globular proteins include enzymes, transport proteins and immunoglobulins.

Fibrous Proteins are structural proteins, generally insoluble in water, consisting of long cable-like structures built entirely of either helical or sheet arrangements. Examples include hair keratin, silk fibroin, elastin, reticulin and collagen. The native state of a protein is its biologically active form.

Intermediate Type Proteins are fibrous, but soluble e.g. fibrinogen which forms insoluble fibrin when blood clots.

Simple and Conjugated Proteins

Simple Proteins: Only amino acids form their structure

Conjugated Proteins: Complex compounds consisting of globular proteins and tightly bound non-protein material; the non-protein material is called prosthetic group. Proteins based on prosthetic groups are listed in Table 1.2.

Table 1.2 Prosthetic group of Conjugated Proteins

Name	Prosthetic Group	Location
Phosphoprotein	Phosphoric acid	Casein of milk, vitellin of egg yolk
Glycoprotein	Carbohydrate	Membrane structure and cell surface receptors
Nucleoprotein	Nucleic acid	Component of viruses, chromosomes (e.g. Histone proteins)

Name	Prosthetic Group	Location
Chromoprotein	Pigment	Haemoglobin – haem (iron containing pigment), Phytochrome (plant pigment)
Lipoprotein	Lipid	Membrane structure. Lipid transported in blood as lipoprotein (e.g. Chylomicrons)
Flavoprotein	FAD (Flavine adenine dinucleotide)	Important in electron transport chain in respiration
Metalloprotein	Metals like Ca, Mo, Zn	Nitrate reductasc (containing Mo), the enzyme in plants which converts nitrate to nitrite

Structural Proteins, Enzymes, Antibodies, Hormones Etc.

Table 1.3   Protein classification According to Function

Type	Examples	Occurrence/function
Structural	Collagen	Component of connective tissue, bone, tendon, cartilage
Enzymes	Trypsin	Pancreatic enzyme; Catalytic hydrolysis of proteins
Hormones	Insulin and glucagons	Help to regulate glucose metabolism
Respiratory pigment	Haemoglobin	Transports oxygen in vertebrate blood
Transport	Serum albumin	Transport of fatty acids and lipids in blood

Type	Examples	Occurrence/function
Protective	Antibodies	Form complexes with foreign proteins
Contractile	Myosin	Moving filaments in myofibrils of muscle
Storage	Casein	Milk protein
Toxins	Diphtheria toxin	Toxin made by diphtheria bacteria

PROTEIN DENATURATION AND RENATURATION

The process of protein denaturation results in the loss of biological activity, decrease in aqueous solubility and increased susceptibility to proteolytic degradation. It can be brought about by heat and by treatment with reagents such as acids and alkalis, detergents, urea, organic solvents and heavy-metal cations such as mercury and lead. It is associated with the loss of organized (tertiary) three-dimensional structure and exposure to the aqueous environment of numerous hydrophobic groups previously located within the folded structure.

Sometimes, a protein will spontaneously refold into its original structure after denaturation, provided the conditions are suitable. This is called renaturation.

PROTEIN PURIFICATION

The purification of a protein from a cell and tissue homogenate typically containing 10000-20000 different proteins seems a frightening task. However, in practice, on average, only four different fractionation steps are needed to purify a given protein. The reason for purifying a protein is normally to provide material for structural or functional studies and the final degree of purity required depends on the purpose for which the protein will be used. Theoretically, a protein is pure when a sample contains only a single protein species, although, in practice, it is more or less impossible to achieve 100%

purity. Many studies on proteins can be carried out on samples that contain as much as 5-10% or more contamination with other proteins. Unavoidably, there will be loss of proteins in each purification step.

The degree of protein purity required depends on the purpose for which the protein is needed e.g. a 90% pure protein is sufficient for amino acid sequence determination studies as long as the sequence is analysed quantitatively to ensure that the deduced sequence does not arise from a contaminant protein. Immunization of a rodent to provide spleen cells for monoclonal antibody production can be carried out with a sample that is considerably less than 50% pure. For studies on enzyme kinetics, a relatively impure sample can be used provided it does not contain any competing activities. On the other hand, for raising a monospecific polyclonal antibody in an animal, it is necessary to have a highly purified protein as antigen; otherwise, immunogenic contaminating proteins will give rise to additional antibodies. Equally, proteins that are used for therapeutic purpose must be extremely pure to satisfy regulatory requirements.

DETERMINATION OF PROTEIN CONCENTRATION

The need to determine protein concentration in solution is a routine requirement during protein purification. The only truly accurate method for determining protein concentration is to acid hydrolyse a portion of the sample and then carry out amino acid analysis on the hydrolysate. However, this is relatively time-consuming, particularly if multiple samples are to be analysed. But, in practice, pure accuracy is not necessary; hence, other quicker methods that give a reasonably accurate assessment of protein concentrations of a solution are used. Most of these are colorimetric methods, where a portion of the protein solution is reacted with a reagent that produces a coloured product. The amount of this coloured product is then measured spectrophotometrically and the amount of colour relates to the amount of protein present by appropriate calibration. However, none of these methods is absolute, since the development of colour is often at least partly dependent on the amino acid composition of the protein(s). The presence of prosthetic groups (e.g. carbohydrate) also influences colorimetric assays. Many workers prepare a standard calibration curve using bovine serum albumin (BSA) because of its low cost, high purity and ready availability. However, it should be understood that, since the amino acid composition of BSA will differ from the composition of the sample being tested, any concentration values deduced from the calibration graph can only be approximate.

Ultraviolet Absorption

The aromatic amino acid residues tyrosine and tryptophan in a protein exhibit an absorption maximum at a wavelength of 280 nm. However, for most proteins, the extinction coefficient lies in the range 0.4-1.5; so for a solution with a complex mixture of proteins with an absorbance of 1.0 at 280 nm (A_{280}), using a 1cm path length, has a protein concentration of approximately 1mgcm^{-3}. The method is relatively sensitive, being able to measure protein concentrations as low as 10 µgcm^{-3}. Unlike colorimetric methods, UV absorption is non-destructive, i.e. after measurement, the sample in the cuvette can be recovered and used for further study. However, the method is subject to interference by the presence of other compounds that absorb at 280 nm. Nucleic acids fall into this category having an absorbance as much as 10 times that of protein at this wavelength. Hence, the presence of only a small percentage of nucleic acid can greatly influence the absorbance at this wavelength. However, if the absorbance (A) at 280 and 260 nm wavelengths are measured, it is possible to apply a correction factor:

$$\text{Protein (mg cm}^{-3}) = 1.55\, A_{280} - 0.76\, A_{260}$$

The great advantage of this protein assay is that it can be measured continuously, e.g. in chromatographic column effluents. Even greater sensitivity can be obtained by measuring the absorbance of ultraviolet light by peptide bonds at 210nm.

Lowry Method (Lowry et al., 1951)

Lowry method (also called as Folin-Ciocalteau) has been the most commonly used one for determining protein concentration. Lowry method is reasonably sensitive, detecting up to 10 µgcm^{-3} of protein, and the sensitivity is moderately constant from one protein to another.

In the presence of alkali, protein molecules form a complex with copper ions and amino acids containing phenolic hydroxyl groups such as tyrosine and tryptophan present in the Cu-protein complex react with Folin-Ciocalteau phenol reagent to give blue color due to the reduction of phosphomolybdate. The intensity of the color is proportional to the concentration of protein. When the Folin reagent which is a mixture of sodium tungstate, molybdate and phosphate, together with a copper sulphate solution, is added to a protein solution, a blue-purple colour is produced which can be quantified by its absorbance at 660 nm. The method is based on the Biuret reaction, where the peptide bonds of proteins react with Cu^{2+} under alkaline conditions producing Cu^+, which reacts with the Folin reagent while in the Folin-Ciocalteau reaction, it involves the reduction of phosphomolybdotungstate to hetero-polymolybdenum blue by the copper-catalysed oxidation of aromatic amino acids. The resultant strong blue colour is therefore partly dependent on the tyrosine and tryptophan content of the protein sample.

Reagents

1. Reagent A: 2% (w/v) sodium carbonate in 0.1N NaOH 1.35% sodium potassium tartrate

2. Reagent B: Dissolve 5.0 mg of cupric sulfate in 1.0 mL of 1.35% sodium-potassium tartrate (w/v), just prior to use.

3. Alkaline copper reagent (Reagent C): Prepare freshly, at the time of protein assay, by mixing 50 mL of reagent A with 1mL of Reagent B (v/v).

4. Folin-Ciocalteau phenol reagent 1N: Commercially available 2N Folin-phenol reagent is diluted to 1N by adding equal volume of water (v/v).

5. Bovine serum albumin (BSA): 12.5 mg of BSA in 50 mL of 0.1N NaOH. The standard solution contains 250 µg/ mL of BSA.

Procedure

Suitable aliquots of protein solution (0.01 mL) are taken in test tubes. The volume in each tube is made up to 1.0 mL with water. 5.0 mL of alkaline copper reagent is added, mixed and allowed to stand for 10 min at room temperature (25°-30°C). 0.5 mL of 1N Folin-phenol reagent is added to each tube, shaken well and kept for incubation for 20 min at room temperature (25°-30°C). The intensity of the blue color developed is read at 660 nm against a reagent blank containing all the reagents except the protein solution. Amount of protein present is calculated by referring to the standard graph. BSA protein in the range of 20 to 100 µg/mL is used to prepare the standard curve and can be used for estimating the protein content of test samples. The standard graph is drawn by plotting the concentration of the standard solution on the X-axis and the optical density (OD) on the Y-axis.

$$\text{Protein Concentration} = \frac{\text{OD of unknown x Standard Concentration}}{\text{OD of known}}$$

Bicinchoninic Acid Method

This method is similar to the Lowry method in that it also depends on the conversion of Cu^{2+} to Cu^+ under alkaline conditions. Cu^+ is then detected by reaction with bicinchoninic acid (BCA) to give an intense purple colour with an absorbance maximum at 562 nm. This method is more sensitive than the Lowry method, being able to detect upto 0.5 µg protein cm^{-3}. It is generally more tolerant for the presence of compounds that interfere with the Lowry assay and hence, there is an increasing popularity of this method.

Bradford Method (Bradford, 1976)

This method is based on the binding of Coomassie Brilliant Blue dye to the protein. This makes the choice of a standard difficult. In addition, many proteins will not dissolve properly in the acidic reaction medium. At low pH, the free dye has absorption maxima at 470 and 650 nm, but when bound to protein, it has an absorption maximum at 595 nm. The practical advantages of the method are that the reagent is simple to prepare and that the colour develops rapidly and is stable. Although it is sensitive down to 20 µg protein, it is only a relative method, as the amount of dye binding appears to vary with the content of the basic amino acids arginine and lysine in the protein. Coomassie Brilliant Blue (CBB) dye solution is prepared by dissolving 0.01% CBB (G-250) in 50 mL of 95% ethanol, and to this, 100 mL of orthophosphoric acid (85% w/v) is added and made up to 1 L with double distilled water. This is filtered through Whatman No.1 filter paper and stored in a brown bottle.

Procedure

To 0.1 mL of sample containing 1-10 µg of protein, 1.0 mL of CBB dye is added and mixed well. CBB binds to the aromatic amino acid residues in the protein and gives a blue colored complex which can be measured at 595 nm after 15 min against dye solution as a reference. The concentration of protein can be estimated from the standard graph using BSA standard as described above.

Kjeldahl Method

This is a general chemical method for determining the nitrogen content of any compound. It is not normally used for the analysis of purified proteins or for monitoring column fractions but, it is frequently used for analyzing complex solid samples and microbiological samples for protein content. The sample is digested by boiling with concentrated sulphuric acid in the presence of sodium sulphate (to raise the boiling point) and a copper and/or selenium catalyst. The digestion converts all the organic nitrogen to ammonia, which is trapped as ammonium sulphate. Completion of the digestion stage is generally recognized by the formation of a clear solution. The ammonia is released by the addition of excess sodium hydroxide and removed by steam distillation in a Markham still. It is collected in boric acid and titrated with standard hydrochloric acid using methyl red-methylene blue as indicator. It is possible to carry out the analysis automatically in an autokjeldahl apparatus. Alternatively, a selective ammonium ion electrode may be used to directly determine the content of ammonium ion in the digest. Although Kjeldahl analysis is a precise and reproducible method for the determination of nitrogen, the determination of the protein content of the original sample is complicated by the variation of the nitrogen content of individual proteins and by the presence of nitrogen in contaminants such as DNA. In practice, the nitrogen content of proteins is generally assumed to be 16% by weight.

References

- Berg JM, Tymoczko JL, Stryer L. (2012). In: *Biochemistry*, 7th edition. WH Freeman and Company Inc., New York.

- Bradford MM. (1976). A rapid and sensitive method for the quantitation of microgram quantities of protein utilizing the principle of protein-dye binding. *Anal Biochem* 7: 248- 254.

- Lowry OH, Rosebrough NJ, Farr AL, Randall RJ. (1951). Protein measurement with the Folin phenol reagent. *J Biol Chem.* 193: 265-275.

- Nelson DL, Cox MM. (2017). In: *Lehninger Principles of Biochemistry*, 7th edition. WH Freeman and Company Inc., New York.

CHAPTER 2

ENZYMES- FUNDAMENTALS

Enzymes are catalysts of biological origin and they are characterized by their extraordinary specificity and reactivity in living systems. Enzymes are found in all living organisms where they catalyse and regulate innumerable chemical reactions. Further, researches on the biochemistry of living organisms have demonstrated the importance of enzymes and enzymology has become one of the present day essential sciences.

PROPERTIES OF ENZYMES

Enzymes possess the following major properties:

i.   They are biocatalysts

ii. All are globular proteins

iii. Being proteins, they are coded by DNA

iv. Their presence does not alter the nature of properties of the end product(s) of the reaction

v.   They are very efficient.In other words,a very small amount of the catalyst brings about the change in a large amount of substrate

vi. They are highly specific

vii. The catalysed reaction is reversible and some ATP dependent reactions are irreversible

viii. Their activity is affected by pH, temperature, substrate concentration and time

ix. Enzymes lower the activation energy of the reaction they catalyse

x.   Enzymes possess active sites where the reaction takes place. The sites have specific shapes.

CLASSIFICATION OF ENZYMES

In recent years, the rapid growth in the science of enzymology and the great increase in the number of enzymes have given rise to many difficulties in terminology. The naming of enzymes by individual workers had proven far from satisfactory in practice. In many cases, the same enzyme became known by several different names, while there were cases in which the same name was given to different enzymes. Many of the names conveyed no idea of the nature of the reactions catalysed.

In view of this state of affairs, an enzyme commission (EC) was appointed by the International Union of Biochemistry and its report published in 1964 and updated in 1972, 1978, 1984 and 1992, forms the basis of the present accepted system. The Enzyme Commission has given a rational classification of enzymes based on reaction types and reaction mechanisms. The chemical reaction catalysed is the specific property which distinguished one enzyme from another and it is logical to use it as the basis for the classification and naming of enzymes. The enzymes are divided into groups on the basis of the type of reaction catalysed and this, together with the name of the substrate, provides a basis for naming individual enzyme.

Difference between Trivial and Systematic Nomenclature:

It was recommended by the Enzyme Commission that there should be two nomenclatures for enzymes, one working or trivial and the other one systematic.

The trivial name will be(i) sufficiently short for general use (ii) not necessarily very exact or systematic and (iii) in many cases, the name already in current use e.g. : trypsin, pepsin and chymotrypsin.

The systematic name of an enzyme will (i) be formed in accordance with definite rules (ii) identify the enzyme precisely (iii) show the action of the enzymes as exactly as possible and (iv) include the name of the substrate.

According to systematic nomenclature, any enzyme is indicated by a code number, unique for each enzyme. This number contains four elements, separated by points and arranged on these principles.

All enzymes are classified into six major categories.

1. *Oxidoreductases*

These enzymes catalyse the removal of H_2 (dehydrogenases) or addition of O_2 (oxygenases). They catalyse the oxidoreduction of

$$\text{>CHOH,} \quad \text{>CH-CH<,} \quad \text{—CHNH}_2 \text{ and others.}$$

eg: L-lactate: NAD^+ oxidoreductase (E.C. 1.1.1.27) (trivial name: lactate dehydrogenase) catalyses:

$$H_3C-\underset{\underset{OH}{|}}{CH}-COO^- + NAD^+ \rightleftharpoons H_3C-\underset{\underset{O}{\|}}{C}-COO^- + NADH + H^+$$

Lactate   Pyruvate

It is the alcohol group of lactate which is involved in the reaction and this is indicated in the above equation.

2. *Transferases*

These enzymes catalyse reactions of the type:

$$AX + B \rightleftharpoons BX + A$$

and they catalyse transfer of groups which are not free during reaction from one substrate to another.

Examples of transferred groups: $-CH_3$, $-CH_2OH$, $-NH_2$

eg: ATP: D-hexose-6-phosphotransferase (E.C. 2.7.1.1) (trivial name: hexokinase) catalyses:

$$C_5H_9O_5.CH_2OH + ATP \rightleftharpoons C_5H_9O_5.CH_2OPO_3^{2-} + ADP$$

D-Hexose $\qquad\qquad$ D-Hexose-6-phosphate

This enzyme will transfer phosphate to a variety of D-hexoses.

3. *Hydrolases*

These enzymes catalyse the splitting of compounds by addition of water across various bonds e.g. peptides, glycoside, ester.

eg: Orthophosphoric monoester phosphohydrolase (E.C.3.1.3.1) (trivial name: alkaline phosphatase) catalyses:

$$R-O-PO_2^--O^- + H_2O \rightleftharpoons R-OH + HO-PO_2^--O^-$$

Organic phosphate $\qquad\qquad\qquad\qquad$ Inorganic phosphate

Alkaline phosphatases are non-specific and act on a variety of substrates at alkaline pH.

Of the hydrolases, proteases are mainly used in the leather industry. Again, proteases are divided into three groups depending upon the origin viz. plant, animal and microbial sources.

4. *Lyases*

These enzymes catalyse the non-hydrolytic addition or removal of groups which are free during reaction. Examples of cleaved bonds: C-C in decarboxylation, C-O in removal of water, C- N and C-S.

eg: L-histidine carboxy-lyase (E.C.4.1.1.22) (trivial name: histidine decarboxylase) catalyses:

Histidine   Histamine

5. *Isomerases*

These enzymes catalyse different types of isomerization e.g.: cis-trans, keto-enol, racemisation or epimerization.

eg: Alanine racemase (E.C.5.1.1.1) catalyses:

$$\text{L - alanine} \rightleftharpoons \text{D - alanine}$$

Alanine
racemase

6. *Ligases*

These enzymes catalyse the condensation of two molecules coupled with the breakdown of a pyrophosphate bond from a nucleotide triphosphate.

The reactions are represented as:

$$X + Y + ATP \rightleftharpoons X - Y + ADP + P_i$$

or

$$X + Y + ATP \rightleftharpoons X - Y + AMP + (PP)_i$$

eg: L-glutamate: ammonia ligase (E.C.6.3.1.2) (trivial name: glutamine synthetase) catalyses:

L-Glutamate $+$ NH_3 $+$ ATP $\rightleftharpoons$ L-Glutamine $+$ ADP $+$ P_i

SPECIFICITY

One of the remarkable properties of enzymes is that they differ from chemical catalysts most strikingly in their specificity. Enzyme specificity is tested by utilizing a number of compounds in which the structure is varied systematically. Enzymes are highly specific both in the reaction catalysed and in their choice of reactants, which are called 'substrates'.

Absolute and Relative Specificity

If an enzyme acts only on one or two substrates, it is absolutely specific. A typical example of absolute specificity is the action of urease on urea to form ammonia and carbon dioxide. Another type of specificity known as relative specificity is seen among the peptidases that catalyse the hydrolysis of peptide bonds in protein. Even though all peptidases catalyse the hydrolysis of peptide bonds, their specificities are different since each can recognize and hydrolyse peptide bonds in different positions in a protein. Chymotrypsin recognises and hydrolyses only those peptide bonds next to phenylalanine and tyrosine. Trypsin acts only on peptide bonds next to basic amino acids and its action is seen below:

Substrate Specificity

Trypsin acts only at peptide, amide or ester bonds involving arginine or lysine residues. α-Benzoyl-L-argininamide or α-Benzoyl-L-lysinamide serves as model substrate. The specificity for reaction of the cationic group of the substrate with an aspartyl group of the enzyme is so exact that deamination will not occur with compounds in which the cationic group is displaced by one CH_2 group from the amide bond. Neither the shorter chain of α-Benzoyl-L-ornithinamide nor the longer chain of α- Benzoyl- L- homoargininamide provides a suitable structure for reaction with trypsin as shown in Figure 2.1.

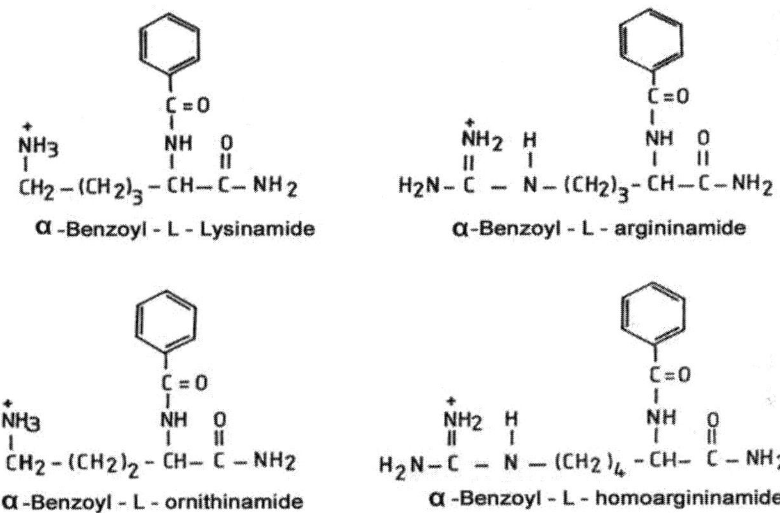

Figure 2.1 Specificity of trypsin

The specificity of some proteases showing their major sites of action are given in Table 2.1.

Table 2.1 Specificity of some proteases showing their major sites of action.

Enzyme	Source	Major site of action
Trypsin	Pancreas	Arginine (Arg), Lysine (Lys)
Chymotrypsin	Pancreas	Tryptophan (Trp), Phenylalanine(Phe), Tyrosine (Tyr)
Pepsin	Gastric mucosa	Trp, Phe, Tyr, Methionine (Met),Leucine(Leu)
Papain	Papaya fruit	Arg, Lys, Glycine (Gly)

Active Site

The active site of an enzyme is the region that binds the substrate (and the prosthetic group, if any) and contributes the residues that directly participate in the making and breaking of bonds. These residues are called as catalytic groups. The active site is a three dimensional entity. The active site of an enzyme is not a point, a line, or even a plane. It is an intricate three dimensional form made up of groups that come from different parts of the linear amino acid sequence e.g, in lysozyme, the important groups in the active site are contributed by residues numbered 35, 52, 62, 63, 101 in the linear sequence of 129 amino acids.

The specificity of binding depends precisely on the defined arrangement of atoms in an active site. A substrate must have a matching shape to fit into the site. Emil Fischer's Lock and Key model suggests that an enzyme like a lock, can only be fitted by a substrate (Key) of precisely the correct shape. This view is modified in the Induced Fit Model of Koshland where an enzyme changes its conformation and binds the substrate in such a way as to make the binding more complete (Figure 2.2a and 2.2b). The substrate binds to the enzyme because the arrangement of charged groups on the enzyme exactly compliments the charged groups on the substrate, or the hydrophobic parts of the substrate exactly fit hydrophobic grooves or pockets in the enzyme.

(a) **Lock and Key Model:** The substrate and enzyme active site have complementary shapes

(b) **Induced Fit Model:** The enzyme active site forms a complementary shape to the substrate after binding

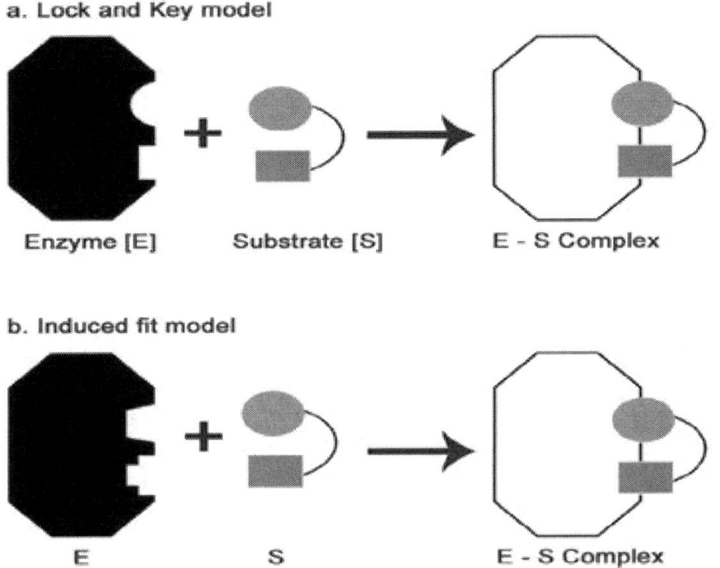

Figure 2.2 Schematic view of Lock and Key Model and Induced fit model

Chymotrypsin is irreversibly inactivated by reaction with certain phosphorus derivatives e.g Diisopropyl fluorophosphate (DFP). A derivative is formed with the specific serine residue (No.195) of the enzyme which is essential for the activity. There are about 28 serines residues but only one combines with DFP. This shows that only a small well defined part of chymotrypsin is responsible for its activity and the active serine (No.195) forms a part of the 'active site' of the enzyme. Considerable evidence also implicates a histidine residue (No.57) in the catalytic activity of chymotrypsin. Treatment of chymotrypsin with L-1(p-toluenesulfonyl) amido-2-phenylethyl chloromethyl ketone (TPCK) results in loss of activity and formation of the imidazole derivative substituted at the third position of histidine residue (No.57). From these examples, it is seen that serine (No.195) and histidine (No.57) form part of the 'active site' of chymotrypsin. The reagents commonly used to detect specific groups and the groups attacked are shown in Table 2.2.

Table 2.2   Compounds commonly used to detect specific groups on the active site

Compound formula	Compound	Main reactive groups
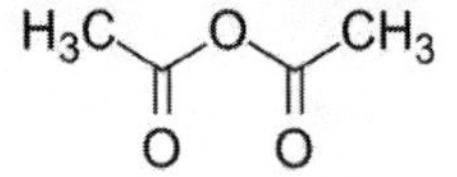 H_3C—O—CH_3 (acetic anhydride structure)	Acetic anhydride	Lys-(NH_2), Ser (-OH), Tyr (-OH)
$I-CH_2-COOH$	Iodoacetic acid	Cys (-SH), His (-NH)
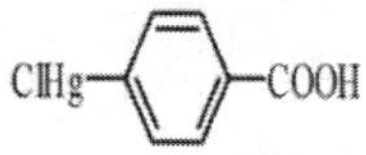 $ClHg$—⬡—$COOH$	p-chloromercury benzoate	Cys (-SH)
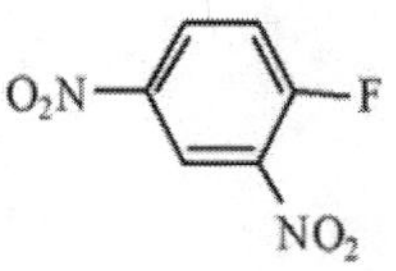 O_2N—⬡—F (with NO_2)	2, 4 - dinitrofluorobenzene	Lys (-NH_2), Tyr (-OH), His (-NH)
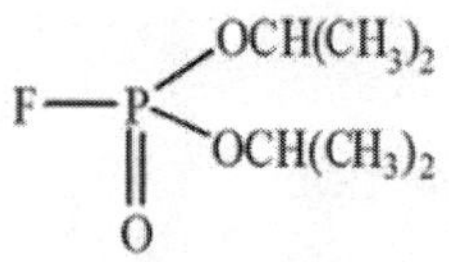 F—P(=O)($OCH(CH_3)_2$)$_2$	Diisopropylfluoro phosphate (DFP)	Ser(-OH) on active centre of chymotrypsin
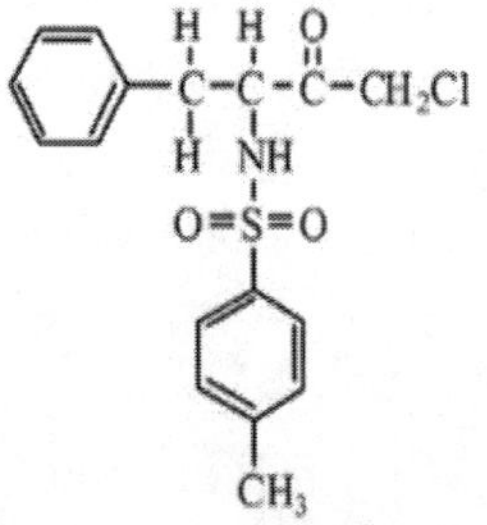 (TPCK structure)	L-1(p-toluenesulfonyl) amido-2-phenylethyl chloromethylketone (TPCK)	His (-NH) on active centre of chymotrypsin

Isoenzymes

Isoenzymes are multiple forms of an enzyme with the same substrate specificity. The multiple forms are due to the genetic differences in their primary structures. They catalyse the same chemical reaction but differ in some of their physico-chemical properties. They can be separated using the techniques of electrophoresis and chromatography. The physico-chemical properties of these enzymes often affect their catalytic activity and they also differ in properties such as their K_m, sensitivity to heat and effect of inhibitors. Lactate dehydrogenase (LDH) is a typical example of an isoenzyme which catalyses the reaction:

$$H_3C-\underset{\underset{OH}{|}}{CH}-COO^- + NAD^+ \rightleftharpoons H_3C-\underset{\underset{O}{\|}}{C}-COO^- + NADH + H^+$$

Lactate   **Pyruvate**

The enzyme, as found in many species, is a tetramer of molecular weight about 1, 40,000 daltons. Although each subunit has a molecular weight of about 35,000 daltons, two types of different amino acid composition are found within each species. They are the M-form, which predominates in skeletal muscle and the H-form, which is a predominant subunit in the heart. The two types of subunits are produced by separate genes. Each monomer is catalytically inactive but it can combine with others of the same or different type to produce the active tetrameric enzymes. Five isoenzymes of LDH can exist viz. H_4 (LDH_1), H_3M (LDH_2), H_2M_2 (LDH_3), HM_3 (LDH_4), and M_4 (LDH_5). Although they catalyse the same reaction, they do so with different characteristics. The properties of H_3M, H_2M_2 and HM_3 are intermediate between those of H_4 and M_4.

COENZYMES AND ACTIVATORS

In many cases, if an enzyme is mixed with its substrate under the appropriate conditions, either no catalysis occurs or there is only a slight activity. This is due to the absence of a coenzyme or activator.

Coenzymes: These are low molecular weight organic compounds which are actively involved in the catalysis. They often act as acceptors or donors of specific chemical groups. Nicotinamide adenine dinucleotide (NAD), for example, accepts and donates hydrogen atoms and is a coenzyme for many dehydrogenases. Coenzyme is the name given to soluble cofactors while the term 'prosthetic' group is reserved for coenzymes that are firmly attached to the protein. Some examples of coenzymes and their involvement in enzymatic reactions are shown in Table 2.3.

Table 2.3 Coenzymes and the enzymatic reactions in which they are involved

Coenzyme	Enzyme
Nicotinamide adenine dinucleotide (NAD$^+$) and Nicotinamide adenine dinucleotide phosphate (NADP$^+$)	Alcohol dehydrogenase
Flavin adenine dinucleotide (FAD)	Glucose oxidase
Coenzyme A (CoA. SH)	Pyruvate dehydrogenase
Pyridoxal phosphate	Aspartate transaminase
Biotin	Acetyl CoA carboxylase
Tetrahydrofolate	Formate tetrahydrofolate ligase
Coenzyme B$_{12}$	Methyl malonyl-CoA mutase
Adenosine triphosphate	Pyruvate kinase

Activators: Non-specific substances which are necessary for the activity of an enzyme or which activate a precursor of an enzyme are often called activators. In most of the enzymatic reactions, metallic ions perform this role. In metal activated enzymatic reactions, purified enzymes may have to be activated by the addition of metal ions. It has already been shown that ternary complexes are formed between an enzyme, metal ion and substrate. The involvement of metal ions in enzymes may be investigated by nuclear magnetic resonance (NMR), electron spin resonance (ESR) and proton relaxation rate (PRR) enhancement techniques. Metalloenzymes are enzymes in which the metal is tightly bound to the enzyme structure and is retained by the enzyme on purification. There is no clear cut distinction between metalloenzymes and metal activated enzymes.

Examples of a few metalloenzymes or metal activated enzymes with their activators are shown in Table 2.4.

Table 2.4 Metalloenzymes/Metal activated enzymes and their activators

Enzyme	Activator
α-amylase & trypsin	Calcium
Creatine kinase and pyruvate kinase	Magnesium
Nitric oxide reductase	Molybdenum and Iron
Carboxy peptidase A & Carbonic anhydrase	Zinc
Arginase and yeast phosphatase	Manganese, Cobalt, Nickel, Iron
Tyrosinase and Ascorbic acid oxidase	Copper

SYNZYMES

A relatively new approach involves synthesizing molecules mimicking the action of natural enzymes due to the incorporation of some features in their active sites. Such artificial enzymes are called as synzymes. Synzymes follow the Michaelis-Menten kinetics just as natural enzymes.

Some synzymes are derivatised proteins. An example is myoglobin which functions as oxygen carrier in muscle. When the group $[Ru(NH_3)_5]^{3+}$ is attached to three surface histidine residues of myoglobin, it starts functioning like an oxidase. It oxidizes ascorbic acid and reduces molecular oxygen. This derivatized myoglobin is almost as effective as the natural ascorbate oxidase.

A different example for the synzyme is that the starting material need not necessarily be a protein. Synzymes with chymotrypsin like characteristics have been obtained based on cyclodextrins, consisting of 6-10 D-glucose units linked head-to-tail in a ring. Glycosidic oxygen atoms and –CH linkages point inwards, creating a hydrophobic environment which can act as a binding pocket. Catalytic activity is provided by the attachment of imidazole, hydroxyl and carboxyl groups, as seen in the active sites of serine proteases. The resulting synzyme resembles chymotrypsin in its esterase activity and it is stable. Cyclodextrins may also be linked to pyridoxyl coenzymes, producing synzymes with transaminase activity.

ABZYMES

Abzymes are antibodies, capable of catalyzing a chemical reaction. One of the major factors governing specificity is the stability of the enzyme-bound transition state which exists during the conversion of enzyme-bound substrate to products. A potential substrate which can form a relatively stable transition-state when bound to the enzyme will be converted to products at an appreciable rate. Transition state analogues are stable compounds which resemble the transition-state compounds thought to be formed as part of a reaction sequence. When an analogue of the supposed transition-state of a particular reaction is injected into a mouse, the immune system will make antibodies against it and some of these may be able to catalyse the reaction in question. These abzymes may be proliferated by the techniques of monoclonal antibody production or by recombinant DNA technology. The first antibody to be commercialized is the abzyme with an aldolase activity. Aldolase abzyme is to be used for the synthesis of Epothilone A, an anticancer compound.

EXTREMOZYMES

It is now known that enzymes can function at temperatures as high as 140°C and as low as below freezing point. Organisms living in these conditions are described as extremophiles and the enzymes that function under these conditions are therefore described as extremozymes. Most of these extremophiles belong to archaea group of bacteria.

Extremozymes are broadly classified as follows:

i. Extremely psychrophilic enzymes, which function at temperatures approaching the freezing point of water; for instance, subtilisin S41 from the psychrophile Bacillus S41 differs from other subtilisins in several respects, suggesting structural differences between these psychrophilic enzymes and those extracted from mesophiles.

ii. Extremely halophilic enzymes function better in salt solutions. For instance, aspartate aminotransferase from *Haloferax mediterranei* is inactivated rapidly at room temperature in low salt solutions, but does not denature even at 78.5°C in 3.3M KCl suggesting that the enzyme is more stable in higher salt concentration. A protease from *Halobacterium halobium* is also similarly affected.

iii. Extremely thermophilic enzymes function at high temperatures and several examples of these enzymes are available now. (DNA polymerase from *Thermus aquaticus* and *Pyrococcus furiosus* used for PCR is one such example).

iv. Extremely barophilic enzymes have been isolated from microbes living in deep sea which are both barophiles and psychrophiles.

FACTORS AFFECTING ENZYMATIC ACTIVITY

Substrate Concentration

The rate of enzyme catalysed reactions increases with increasing substrate concentration but above a certain substrate concentration, the rate of enzyme action ceases to increase (Figure.2.3). The shape of the curve is commonly explained on the basis of catalytically active sites on the enzyme that react with the substrate. Maximum velocity V_{max}, of the reaction is reached when all the sites are occupied by the substrate molecules.

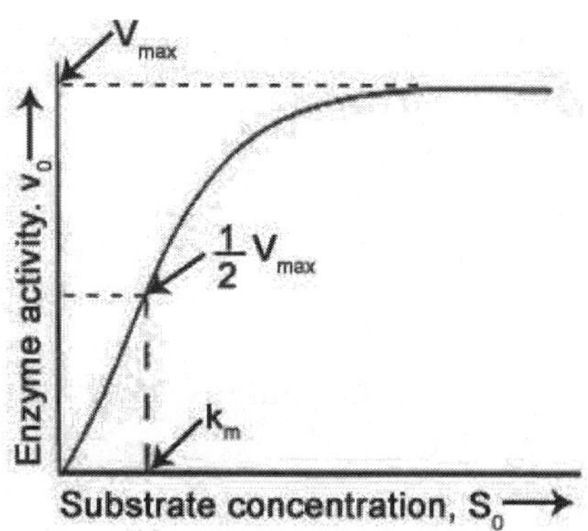

Figure 2.3   Effect of substrate concentration on enzyme activity

Michaelis- Menten Equation   L.Michaelis and M.L.Menten in 1913, postulated the existence of an enzyme-substrate complex as the basis for a theoretical analysis of enzymatic reactions. Let us consider a single substrate enzyme catalysed reaction where there is just one substrate-binding site per enzyme.

$$\text{Enzyme} + \text{Substrate} \underset{k_{-1}}{\overset{k_1}{\rightleftharpoons}} \text{Enzyme} - \text{Substrate complex} \rightarrow \text{Enzyme} + \text{Product}$$
$$\text{E} \qquad \text{S} \qquad\qquad \text{ES} \qquad\qquad\qquad k_2$$

The Michaelis- Menten assumption was that equilibrium between enzyme, substrate and enzyme-substrate complex was instantly set up and maintained and the breakdown of the enzyme-substrate complex to products was negligible to disturb this equilibrium. Briggs and Haldane extended this idea and derived an equation assuming steady-state conditions i.e., the rate of breakdown of the complex is the same as the rate of formation during the period of the measurement. Using this assumption, the equation is written as

$$k_1[E][S] = k_{-1}[ES] + k_2[ES] = [ES](k_{-1} + k_2)$$

Separating the constants from the variables,

$$\frac{[E][S]}{[ES]} = \frac{\left[k_{-1} + k_2\right]}{k_1} = K_m$$

where k_m is another constant.

The total concentration of enzyme present $[E_0]$ must be the sum of the concentration of free $[E] = [E_0] - [ES]$ enzyme $[E]$ and the concentration of bound enzyme $[ES]$. Substituting in the above equation,

$$\frac{([E_0] - [ES])\,[S]}{[ES]} = K_m$$

$$K_m[ES] = ([E_0] - [ES])[S]$$

$$= [E_0][S] - [ES][S]$$

$$[ES][S] + K_m[ES] = [E_0][S]$$

$$[ES]([S] + K_m) = [E_0][S]$$

$$[ES] = \frac{[E_0][S]}{[S] + K_m}$$

The term $[ES]$ governs the rate of formation of products (the overall rate of reaction) according to the relationship:

$$v_0 = k_2[ES]$$

If this is substituted in the above equation,

$$\frac{v_0}{k_2} = \frac{[E_0][S]}{[S] + K_m}$$

$$v_0 = \frac{k_2[E_0][S]}{[S] + K_m}$$

Moreover, when the substrate concentration is very high, all the enzyme is present as the enzyme-substrate complex and the maximum velocity, V_{max}, is reached. Under these conditions,

$$V_{max} = k_2[E_0]$$

Hence,

$$v_0 = \frac{V_{max}[S]}{[S] + K_m}$$

Finally, since the substrate concentration, $[S_0]$, is usually greater than the enzyme concentration,

$$v_0 = \frac{V_{max}[S_0]}{[S_0] + K_m} \text{ at constant } [E_0]$$

This equation has retained the name Michaelis-Menten equation and K_m is called Michaelis constant.

A graph of v_0 against $[S_0]$ will have the form of rectangular hyperbola consistent with experimental findings for many enzyme catalysed reactions. V_{max}, the maximum initial velocity at a particular $[E_0]$ can be obtained from the graph as shown in Figure 4.3. K_m can also be $v_0 = \frac{1}{2}V_{max}$ obtained from the graph.

and, when this is substituted in Michaelis-Menten equation,

$$\frac{V_{max}}{2} = \frac{V_{max}[S_0]}{[S_0] + K_m}$$

$$V_{max}([S_0] + K_m) = 2(V_{max})[S_0]$$

$$K_m = [S_0]$$

Therefore, K_m is the substrate concentration at which an enzyme reaches half the maximal velocity.

Significance of Michaelis- Menten Equation The Michaelis-Menten equation, as modified by Briggs and Haldane, is applicable to many enzyme-catalysed reactions and the constants V_{max} and K_m can be determined using this equation. V_{max} varies with the total concentration of the enzyme present, but K_m is independent of enzyme concentration and is characteristic of the system being investigated. It can, thus, be used to identify a particular enzyme. K_m value of an enzyme varies widely. This means that the substrate concentration needed to saturate the enzyme is different for different enzymes. An enzyme with a large K_m value is one that is saturated at low substrate concentration. For the same enzyme, K_m value changes when different substrates are used. In this context, a large K_m value indicates that the affinity of an enzyme for a particular substrate is low and a small K_m value indicates an inverse relationship. K_m values of some enzymes are given in Table 2.5 and for most enzymes, K_m value lies between 10^{-1} and 10^{-6}M.

Table 2.5 K_m values of some Typical Enzymes

Enzyme	Substrate	K_m (M)
Chymotrypsin	Acetyl-L-tryptophanamide	5×10^{-3}
Lysozyme	Hexa -N-acetylglucosamine	6×10^{-6}
β-galactosidase	Lactose	4×10^{-3}
Trypsin	Benzoyl-L-arginine ethyl ester	5×10^{-5}
Pepsin	Acetyl-L-phenylalanyl Diiodotyrosine	7.5×10^{-5}
Papain	Benzoyl-L-arginine ethyl ester	1.89×10^{-3}
β-Amylase	Amylose	7.3×10^{-5}

Lineweaver – Burk Plot   The graph of the Michaelis-Menten equation showing enzyme activity against substrate concentration is not entirely satisfactory for the determination of V_{max} and K_m (Figure 2.3). Unless there are at least three consistent points on the plateau of the curve at different [S] values, an accurate value of V_{max} and hence of K_m cannot be obtained. The graph, being a curve, cannot be accurately extrapolated upwards from non-saturating values of [S].

Lineweaver and Burk overcame this problem without making further assumptions. The Michaelis- Menten equation was simply inverted as shown below:

$$V_0 = \frac{V_{max}[S_0]}{[S_0]+K_m}$$

$$\frac{1}{V_0} = \frac{[S_0]+K_m}{V_{max}[S_0]}$$

$$= \frac{[S_0]}{V_{max}[S_0]} + \frac{K_m}{V_{max}[S_0]}$$

$$\frac{1}{V_0} = \frac{K_m}{V_{max}}\frac{1}{[S_0]} + \frac{1}{V_{max}} \text{(Lineweaver - Burk equation)}$$

This is of the form y=mx + c, which is the equation for straight line graphs. A plot of y against x has a slope m and intercepts c on the Y-axis. A plot of $1/v_0$ against $1/[S_0]$ (Lineweaver- Burk plot) for systems obeying the Michaelis-Menten equation is shown in Figure 2.4. The graph being linear can be extraploted even if no experiment has been performed at a saturating substrate concentration and from the extrapolated graph, the values of K_m and V_{max} can be determined.

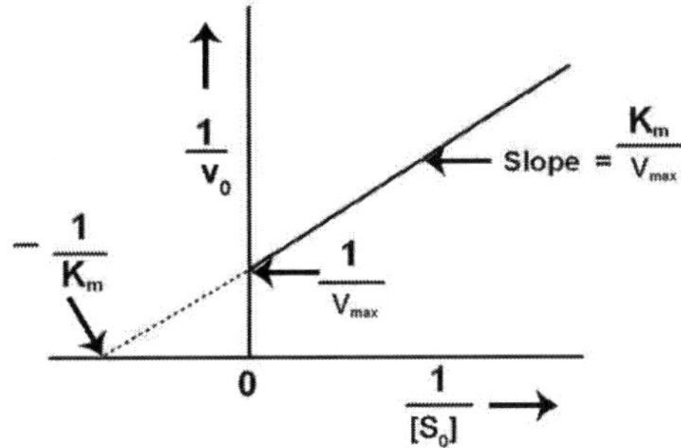

Figure 2.4   Lineweaver-Burk plot

Enzyme Concentration

The relation between enzyme concentration and its activity is demonstrated by maintaining substrate concentration, pH and temperature constant while the concentration of the enzyme solution is varied.

Three different conditions are expressed in Figure 2.5. 'A' indicates normal response showing that the activity varies linearly with enzyme concentration. 'B' shows that some activator is present in the enzyme preparation and hence the reaction does not proceed in a linear way. 'C' shows a condition of substrate exhaustion where there is no increase in reaction velocity once the substrate is depleted.

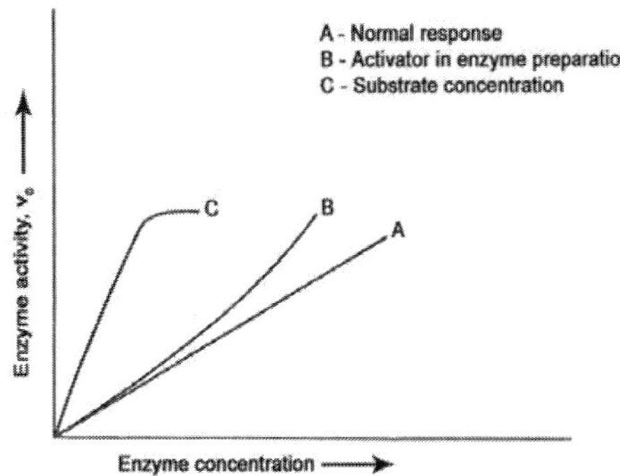

Figure 2.5   Effect of enzyme concentration on activity

Temperature

Enzyme catalysed reactions are similar to other reactions in that the rate is increased by increasing temperatures upto a point. Beyond that temperature, the activity of an enzyme declines sharply as shown in Figure 2.6. As the temperature increases beyond 45° - 50° C, the rate decreases. This decrease is caused by thermal denaturation of the enzyme protein or the inactivation of a thermolabile component in the enzyme system. When thermal energy becomes great enough to cause the rupture of a few bonds, the neighbouring bonds are weakened and the whole molecule becomes denatured. The optimum temperature of enzymes under physiological conditions is close to 40°C. The maximum velocity of enzyme reaction is obtained around the optimum temperature.

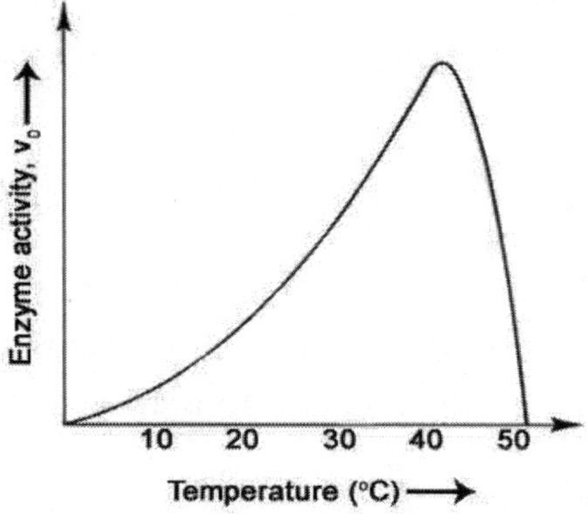

Figure 2.6   Effect of temperature on enzyme activity

pH

When the enzymatic activity is plotted against pH, a bell shaped curve is obtained. This indicates a marked dependence of an enzyme on the pH of the reaction mixture with enzyme activity decreasing rapidly on either side of the optimum pH (Figure 2.7). The pH optimum of an enzyme is dependent upon a number of experimental parameters including temperature, nature of substrate, concentration of buffer, ionic strength of medium and purity of enzyme preparation.

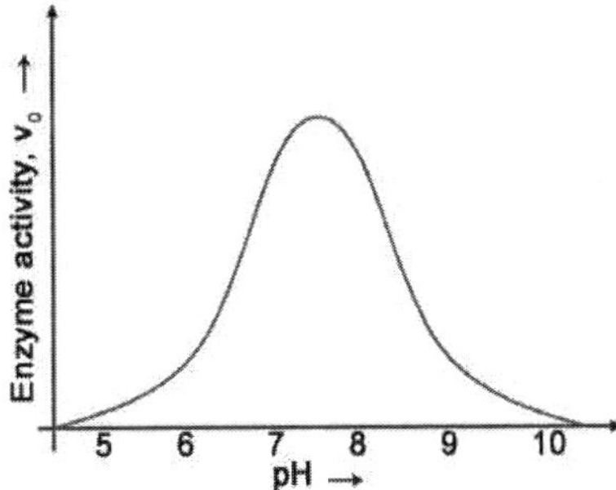

Figure 2.7   Effect of pH on enzyme activity

INHIBITORS

An enzyme inhibitor is a chemical that combines with an enzyme and reduces or eliminates its catalytic activity. The decrease in catalytic activity means that the inhibitor has affected the active site of the enzyme directly, by damaging or physically clogging up the active site. Indirectly, it may change the three-dimensional shape of the entire protein. Some inhibitors act against only one enzyme, while others inactivate a wide spectrum of enzymes. There are two types of inhibitors: (i) Irreversible (ii) Reversible. Irreversible inhibitors cannot be removed from an enzyme by dialysis. Reversible inhibitors bind to an enzyme in a reversible fashion and can be removed by dialysis. Simple dilution can restore full enzyme activity. Reversibility refers to whether or not the enzyme-inhibitor complex can dissociate to regenerate the enzyme.

Irreversible Inhibition

An irreversible inhibitor binds to the active site of the enzyme by an irreversible $E + I \rightarrow EI$ reaction: and hence cannot subsequently dissociate from it. A covalent bond is usually formed between inhibitor and enzyme. The inhibitor may act by preventing enzyme-substrate binding or it may destroy some component of the catalytic site. Irreversible inhibitors effectively reduce the activity of enzyme present.

Many irreversible inhibitors attack –SH groups (in cysteine side chains) which are often found at the active sites of enzymes. Important examples are alkylating agents such as iodoacetic acid and iodoacetamide which form covalent linkages with the essential -SH groups.

$$\text{E-SH} + \text{I- CH}_2\text{COOH} \rightarrow \text{E-S-CH}_2\text{COOH} + \text{HI}$$
$$\text{(Enzyme)} \quad \text{(Iodoacetic acid)}$$

Another well known group is the organophosphorous compounds which react with essential -OH groups (in serine side chains) of some enzymes. An example is diisopropylfluorophosphate (DFP). It is a nerve poison and it inactivates acetylcholinesterase, an enzyme essential in nerve function.

Irreversible inhibitors are useful in the investigation of the active site of an enzyme. The inhibitor will remain firmly bound to one of the amino acids in the active site of the enzyme and thus act as a marker for easy identification.

Reversible Inhibition

Some of the major types of reversible inhibition are discussed below:

(a) Competitive inhibition (b) Non-Competitive inhibition (c) Uncompetitive inhibition (d) Substrate inhibition and (e) Allosteric inhibition.

Competitive Inhibition   When a compound competes with a substrate for the active site on the enzyme molecule and thereby reduces the catalytic activity of the enzyme, it is known as competitive inhibitor. These compounds are usually structural analogs of the substrates.

$$E + S \overset{1}{\rightleftharpoons} ES \overset{2}{\longrightarrow} E + P$$

Let us consider a typical enzymatic reaction.

Two conditions exist here. If the inhibitor interferes in the first step, it combines with the active centre of the enzyme and brings about a different orientation in the enzyme. When the inhibitor interferes in the second step, it combines with [ES] complex, retarding its dissociation into product and enzyme. This is possible by changing the conformation of enzyme in the [ES] complex.

An important factor in competitive inhibition is the concentration of substrate and inhibitor with respect to each other. If S:I, ratio known as 'affinity ratio', is 50 : 1, then I is able to alter the reaction velocity. If this ratio is > 50, the substrate nullifies the effect of I. Since a higher concentration of substrate will resume the enzymatic activity, competitive inhibition is reversible. At very high substrate concentration, molecules of substrate will outnumber molecules of inhibitor and the effect of the inhibitor will be negligible. Hence, V_{max} for the reaction is unchanged. As the active sites are directly involved, the K_m of the enzyme is altered i.e. increased by the inhibitor.The Lineweaver-Burk plot showing the effect of competitive inhibition is depicted in Figure 2.8.

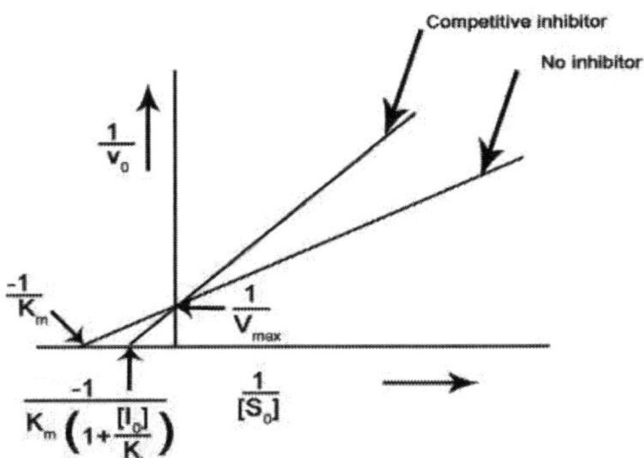

Figure 2.8   Lineweaver - Burk plot showing the effect of competitive inhibition

A typical example of competitive inhibition is cited below. Succinate dehydrogenase (SDH) readily oxidizes succinate to fumarate in the presence of a suitable hydrogen acceptor (A). When inhibitor (I) is present, the rate of reaction is dependent upon the concentration of enzyme, substrate and inhibitor. The structural similarity between the substrate and inhibitor helps the binding of I with E to effect inhibition. In the above reaction, malonate serves as strong inhibitor.

Both malonate and succinate can combine with the active site of the enzyme but no product is formed in the presence of malonate. So, malonate inhibits SDH. Another example of competitive inhibition is the inhibition of folic acid synthesis by sulfanilamide in many microorganisms.

p-aminobenzoic acid   Sulfanilamide

Sulfanilamide, a structural analogue of p-aminobenzoic acid, will block folic acid synthesis and the resulting deficiency of this essential vitamin is fatal to the organism.

Noncompetitive Inhibition   As the name implies, no competition occurs between substrate and inhibitor. In this type, the inhibitor combines rather strongly with a site other than the active site on the enzyme surface (Figure 2.9). Non-competitive inhibitor lowers the maximum velocity (V_{max}) attainable with a given amount of enzyme but does not affect the K_m value. Since I and S may combine at different sites, formation of both EI and EIS is possible. As EIS may break to form the products at a slower rate than ES, the reaction is slowed down but not stopped. The reactions are shown below:

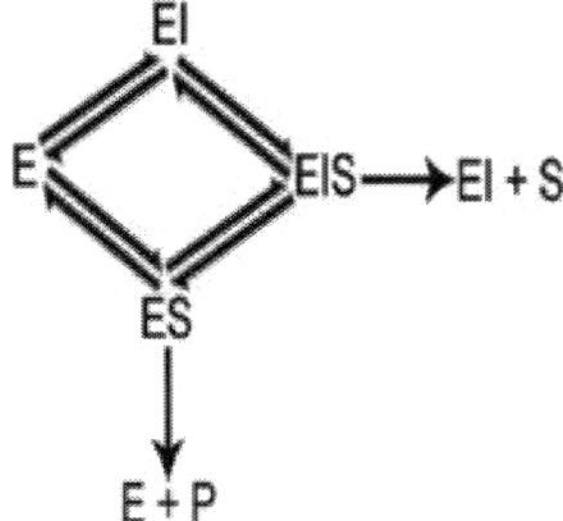

When $1/v_0$ is plotted against $1/[S_0]$ at various concentrations of inhibitor, the results, as shown in (Figure 2.9), are obtained. There has been no significant alteration of the conformation of active site when I is bound.

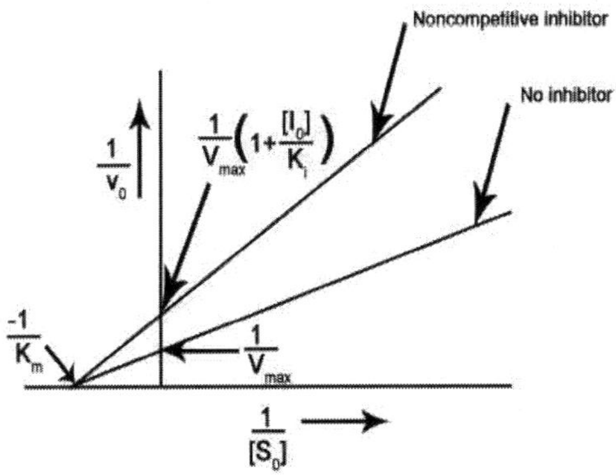

Figure 2.9 Lineweaver - Burk plot showing the effect of simple linear noncompetitive inhibition

Uncompetitive Inhibition Uncompetitive inhibitors bind only to the enzyme substrate complex and not to the free enzyme. Substrate-binding could bring about a change in the conformation of the enzyme and reveal an inhibitor binding site. The inhibitor does not compete with the substrate for the same binding site and hence, the inhibition cannot be overcome by increasing the substrate concentration. Uncompetitive inhibition may be represented as below:

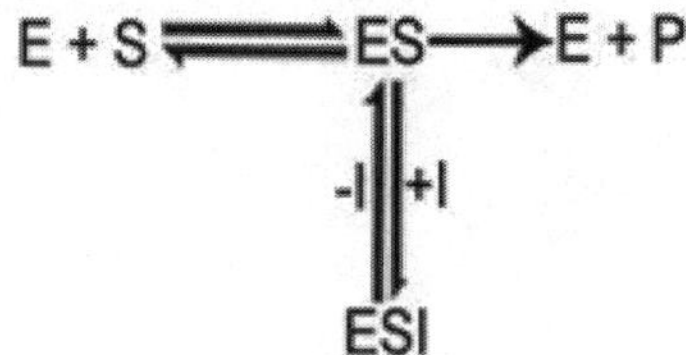

As shown in Figure 2.10, both K_m and V_{max} are altered.

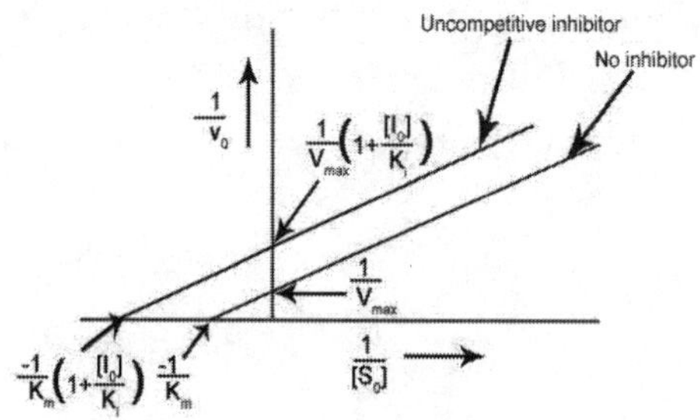

Figure 2.10 Lineweaver-Burk plot showing the effect of uncompetitive inhibition

The diagrammatic representation of competitive, noncompetitive and uncompetitive inhibition is shown in Figure 2.11.

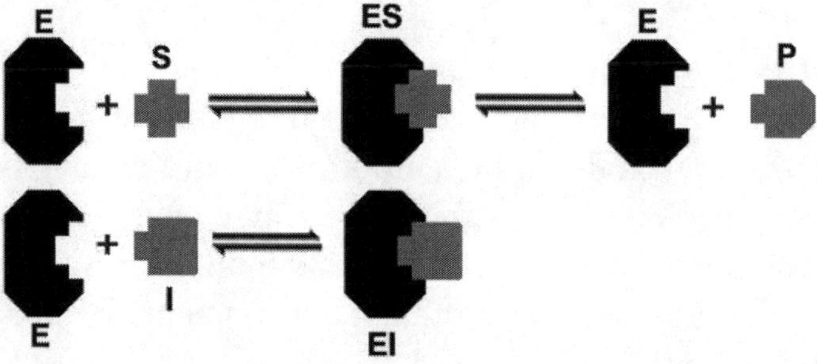

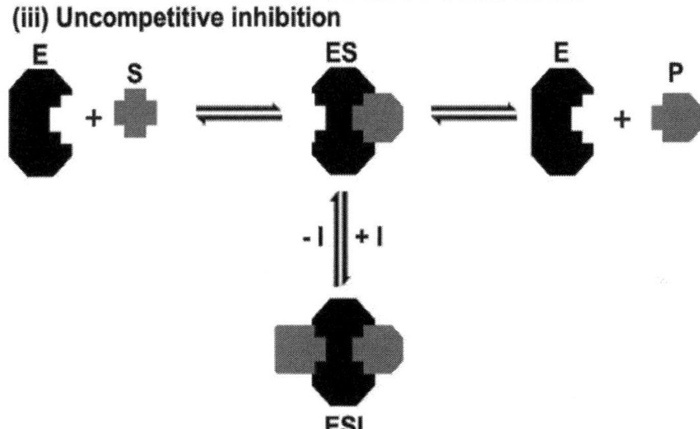

Figure 2.11 Diagrammatic representation of competitive, noncompetitive and uncompetitive inhibition

Substrate Inhibition A characteristic of enzyme catalysed reactions is that for a given enzyme concentration, the initial reaction velocity increases with increasing initial substrate concentration to a limiting value, V_{max}. At still higher substrate concentrations, the initial velocity is sometimes found to be less than the maximum value. It could be concluded that the substrate, at very high concentrations, can inhibit its own conversion to product.

One possible mechanism for substrate inhibition at high substrate concentration could be explained with relevance to succinate dehydrogenase. For a reaction to take place, both carboxyl groups of the substrate have to bind to the enzyme.

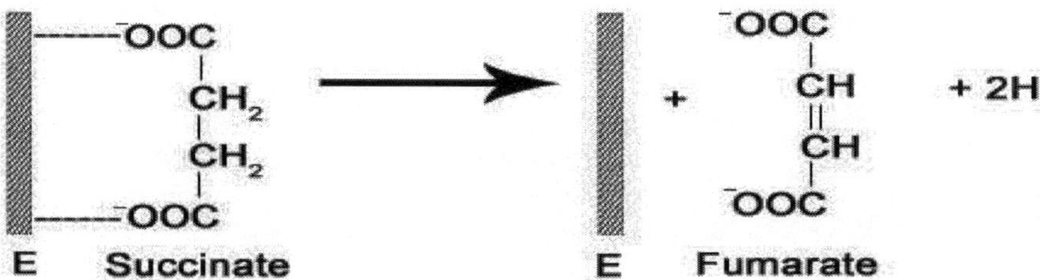

At very high substrate concentrations, there is an increased possibility of carboxyl groups from two separate substrate molecules binding to the same enzyme. In this particular condition, a reaction cannot take place until one of them dissociates away. The mechanism of substrate inhibition could be explained as shown in Figure 2.12.

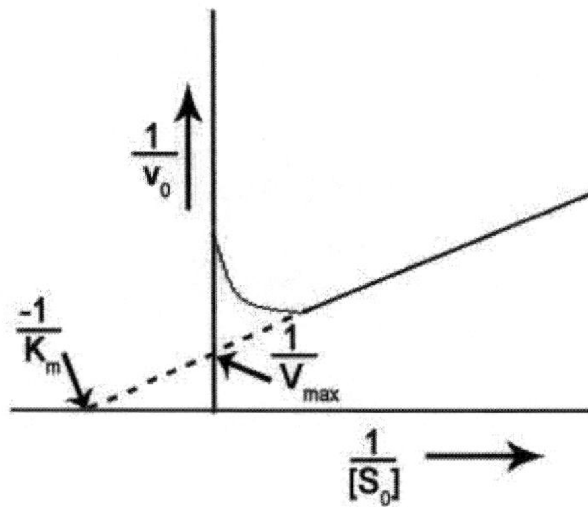

Figure 2.12 Lineweaver-Burk plot showing the effect of substrate inhibition

Substrate inhibition occurs when a molecule of substrate binds to one site on the enzyme and then another molecule of substrate binds to a separate site on the enzyme to form a dead-end complex.

Allosteric Inhibition   An allosteric inhibitor binds to the enzyme at a site other than the substrate binding site. The term "allosteric inhibition" is used when the inhibitor, instead of forming a complex with the enzyme, influences conformational changes which may alter the binding characteristics of the enzyme for the substrate or the subsequent reaction characteristics or both. The Michaelis-Menten plot becomes more sigmoidal (S-shaped) which means that the rate of reaction is reduced at low substrate concentration (Figure 2.13).

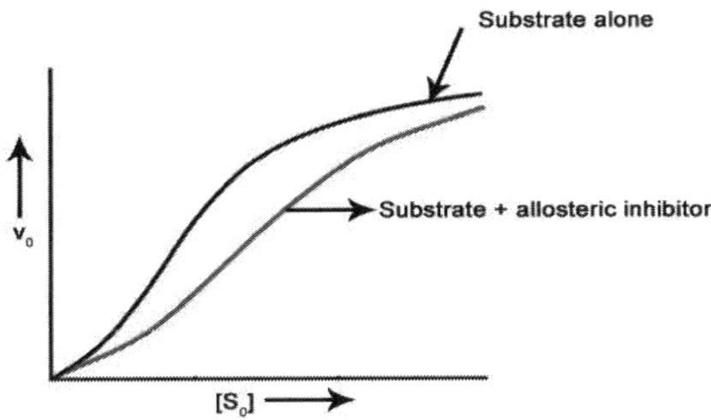

Figure 2.13   Effect of allosteric inhibitor on the binding of a substrate

If the binding characteristics alone are affected, V_{max} will usually remain unchanged and the inhibition pattern could be regarded as competitive. Similarly, other forms of allosteric inhibition, where V_{max} is altered, could be regarded as giving non-competitive or mixed inhibition depending on whether K_m is changed or not. However in most cases, the Michaelis-Menten equation is not obeyed in the presence of allosteric inhibitors. A typical case of Allosteric inhibition is the inhibition of aspartate transcarbamoylase by cytidine triphosphate (CTP).

ENZYME INHIBITORS – APPLICATIONS

Mechanism of enzyme inhibition is important in that (i) it provides an understanding about the possible ways by which metabolic activity could be controlled *in vivo* (ii) It gives an insight into the mechanism by which enzymes promote their catalytic activity and (iii) It allows specific inhibitors to be synthesized and used as therapeutic agents to block key metabolic pathways underlying clinical conditions.Some of the examples of enzyme inhibitors acting as therapeutic agents are shown in Table 2.6.

Table 2.6   Examples of enzyme inhibitors as therapeutic agents

Inhibitor	Enzyme	Application
Aspirin	Cyclooxygenase (COX-1 and COX-2)	Inhibition of prostaglandin and thromboxane synthesis
Mevinolin	Hydroxymethyl glutaryl CoA reductase	Inhibition of cholesterol synthesis
Allopurinol	Xanthine oxidase	Treatment of gout
Disulfiram	Aldehyde dehydrogenase	Treatment of alcoholism
Sulphonamides Trimethoprim	Dihydrofolate synthase	Bacterial infections
Methotrexate	Dihydrofolate synthase	Anticancer effect
Celastatin	Dehydropeptidase-I	Enhancement of antibacterial action
Ritonavir, Saquinavir	HIV protease	HIV therapy

References

- Berg JM, Tymoczko JL, Stryer L. (2012). In: *Biochemistry*, 7th edition. WH Freeman and Company Inc., New York.

- Nelson DL, Cox MM. (2017). In: *Lehninger Principles of Biochemistry*, 7th edition. WH Freeman and Company Inc., New York.

- Palmer T, Bonner PL. (2007). In: *Enzymes: Biochemistry, Biotechnology, Clinical Chemistry*, 2nd Edition. Woodhead Publishing, New Delhi.

CHAPTER 3

PROTEINS IN DIAGNOSIS

INTRODUCTION

Proteins play an important role in clinical diagnosis. Human blood plasma contains proteins at a concentration of approx. 6.0 – 8.3 g/dL. Alterations in the concentration of plasma proteins are observed during disease progression and this provides some important diagnostic clues. This chapter deals with the comprehensive review of major protein markers used in clinical medicine and they are listed in Table 3.1.

Table 3.1 Protein markers in clinical medicine

Auto-immune markers	Cardiac markers	Diabetic markers	Endo-crinology markers	General Biochemical markers	Hematology related markers
• Anti-nuclear antibodies • Anti-sperm antibodies • Anti-phospholipid antibodies/Anti-cardiolipin antibodies • Anti-Beta -2 glycoprotein 1 • Rheumatoid factor • Cyclic citrullinated peptide antibody	• Beta-type natriuretic peptide • Troponins	• Anti -insulin antibody • C - peptide • HbA1c • Insulin	• Luteinizing hormone • Follicle stimulating hormone • Human chorionic gonadotropin • Thyroid stimulating hormone • Parathyroid hormone • Growth hormone • Insulin like growth factor - 1 • Prolactin • Gastrin • Renin • Sex hormone binding globulin	• Adiponectin • Leptin • Total protein • Albumin • Globulin • Micro-albumin • Ferritin	• Hemoglobin • Erythropoietin • Fibrinogen • Factor VIII • von Willebrand factor • Haptoglobulin • Hemopexin

Hepatitis markers	HIV markers	Lipo-protein markers	Nephrology markers	Specific protein markers	Tumor markers
• Anti-HAV • Anti-HCV • HDAg and Anti-HDV • Anti-HEV • HBsAg • HBsAb • HBcAb	• HIV antigen and antibody • CD4 and CD8	• Apolipoprotein A • Apolipoprotein B	• Cystatin C • Kidney injury molecule 1 (KIM-1) • Neutrophil gelatinase associated lipocalin (NGAL) • Fibroblast growth factor 23 (FGF23).	• C reactive protein • hs-CRP • Serum amyloid A protein • Alpha 1 antitrypsin • Anti-streptolysin O • Anti-DNase B • Angiopoietin 2 • Ceruloplasmin • Cryoglobulins • Complements (C3, C4) • Immunoglobulins	• Alpha feto protein • Beta2 microglobulin • Bence Jones protein • Cancer antigen 125 • Cancer antigen 15-3 • Cancer antigen 19-9 • Cancer antigen 72-4 • Carcino embryonic antigen • Prostate specific antigen • CD 20 antigen • Progesterone and Estrogen receptors • Human epidermal growth factor receptor 2 • Thyroglobulin

AUTO-IMMUNE MARKERS

Anti-Nuclear Antibodies (ANA)

Anti-nuclear antibodies (ANA) are auto-antibodies produced by the immune system, which bind to auto-antigens present in the nuclei of mammalian cells. Most of them are IgG but, IgM and IgA are also detected. Presence of ANA in blood could be a marker of an autoimmune process. It is associated with several autoimmune disorders but, it is most commonly seen with systemic lupus erythematosus (SLE). ANA testing may be used along with or followed by other autoantibody tests that target specific substances within cell nuclei, such as anti-dsDNA, anti-centromere, anti-nucleolar, anti-histone and anti-RNA antibodies (Yu et al., 2014).

Anti-Sperm Antibodies (ASA)

Sperm antibodies are produced by the immune system, when sperm comes in contact with blood (both in male and female). In males, this could happen when the testicles are injured or after surgeries or after a prostate gland infection. The testicles normally keep the sperm away from the rest of the body and the immune system. In females, an allergic reaction could be elicited to her partner's semen and sperm antibodies could be produced. These sperm antibodies are produced by the lymphocytes, which are mainly of the IgG, IgA and IgM type. Production of sperm antibodies both in male and female may lead to fertility problems. ASA test is carried out in couples to diagnose infertility issues. Higher the level of antibody-affected sperm found in the semen, lower is the chance of the sperm fertilizing an egg (Naz and Menge, 1994).

Antiphospholipid Antibodies (APA) / Anti-Cardiolipin Antibodies

Phospholipids are structural components of cell membranes and they play a crucial role in blood clotting. In addition to various coagulation co-factors, phospholipids are also critical to platelet function. APA are auto antibodies which are mistakenly produced by the body against its own phospholipids.

The major APA that are diagnosed together usually include anticardiolipin antibodies, lupus coagulant and anti-beta-2 glycoprotein 1 antibodies. When APA are produced, they interfere with the clotting process. They increase the affected person's risk of developing recurrent inappropriate blood clots (thrombi) in arteries and veins, which can lead to strokes and heart attacks. APA are also associated with low platelet counts (thrombocytopenia), with the risk of recurrent miscarriages (especially in the second and third trimester), premature labor and pre-eclampsia. Presence of one or more APA in blood may correlate with autoimmune disorders viz., antiphospholipid syndrome (APS), lupus, rheumatoid arthritis, HIV or rubella infections and tumors of lung, colon etc. APS is an autoimmune disorder characterized by recurrent thrombosis and pregnancy morbidity in the presence of antiphospholipid antibodies (Wong et al., 2004). APS could be primary with no underlying autoimmune disorder or secondary, existing with a diagnosed autoimmune disorder (Bidot et al., 2006).

Cardiolipins are phospholipids which play an important role in blood clotting cascade. They are present in the outermost layer of cells and platelets. Anti-cardiolipin antibodies are autoantibodies, produced by the immune system which targets the body's own cardiolipins. They are determined in the blood when there is frequent thrombotic episode and recurrent miscarriage. Anti-cardiolipin antibodies are diagnosed along with lupus anticoagulant and anti-beta2 glycoprotein 1 antibodies. Presence of this antiphospholipid antibody, along with clinical scenario, may correlate with APS(Lakos et al., 2012).

Anti-Beta -2 Glycoprotein 1 (Anti-Beta-2 GP1)

Beta-2 GP1, also called as apolipoprotein H, is synthesized by hepatocytes, endothelial cells and trophoblasts. Autoantibodies to beta-2 GP1 are produced in individuals afflicted with inappropriate blood clotting. Anti-Beta-2 GP1 is an important APA. Plasma of normal individuals contains very low concentrations of autoantibodies to beta-2 GP1. Biochemical assay detects and measures one or more classes of beta-2 GP1.

Rheumatoid Factor (RF)

Immunoglobulin M (IgM) autoantibodies against the Fc fragment of immunoglobulin G (IgG) are called rheumatoid factors (RFs). These proteins are produced by B cells and they are found circulating in the blood. Their role is unknown in both healthy individuals and in those with rheumatoid arthritis. Blood test for RF is valuable to diagnose rheumatoid arthritis. Approximately 60-80% of individuals with rheumatoid arthritis (RA) have RF present during the course of their disease. However, RF may also be detected in patients with other autoimmune disorders such as SLE, Sjögren syndrome, as well as persistent bacterial, viral, and parasitic infections, endocarditis etc. It may sometimes be seen in those with lung disease, liver disease, and kidney disease, and it can also be found in a small percentage (1-5%) in healthy individuals.

Cyclic Citrullinated Peptide Antibody (CCPA)

CCPA have recently emerged as sensitive and specific serological markers of rheumatoid arthritis (RA). CCPA are autoantibodies produced by the immune system that are directed against cyclic citrullinated peptides (CCP). Citrullination is a post-translational modification of arginine by deimination, physiologically occurring during apoptosis, inflammation or keratinization. The presence of several citrullinated peptides has been demonstrated in the RA synovium (Szekanecz et al., 2008). Though RF is used to diagnose RA, its sensitivity and specificity are low. CCPA are more sensitive and specific and they are more likely to be positive with early RA. In clinical practice, it is recommended to measure CCPA and RF together. Because CCPA alone are only moderately sensitive, combination of the two markers improves diagnostic accuracy, especially in the case of early rheumatoid arthritis (Ingegnoli et al., 2013).

CARDIAC MARKERS

B-type Natriuretic Peptide (BNP) or N–Terminal pro B-type Natriuretic Peptide (NT pro BNP)

BNP is a cardiac neurohormone produced by the cardiac ventricles and it is released as preproBNP peptide of 134 amino acids. It is then cleaved into proBNP (108 amino acids) and as a signal peptide of 26 amino acids. ProBNP is subsequently cleaved into BNP (32 amino acids) and the inactive N-terminal proBNP peptide (NT-proBNP; 76 amino acids). The release of BNP into the circulation is directly proportional to the ventricular expansion as well as volume overload of the ventricles and therefore, it reflects the decompensated state of the ventricles. BNP aids in vasodilation, natriuresis and diuresis and it helps the loading condition of the failing heart. Hence, BNP and NT-proBNP are useful markers for the diagnosis and exclusion of heart failure (Maeda et al., 1998). BNP and NT-pro BNP levels are also elevated in other clinical conditions viz., acute myocardial infarction, atrial fibrillation, cardiac amyloidosis, chronic renal failure, cirrhosis etc and a decreased level is associated with obesity, pericardial constriction, mitral stenosis etc (Daniels and Maisel, 2007).

Troponins

Troponins are protein molecules forming part of cardiac and skeletal muscles. There are 3 types of troponins viz., troponin T, I and C, each with a unique function. Among these, cardiac troponins T (cTnT) and I (cTnI) are specific markers of myocardial injury. Troponins are generally undetectable in healthy patients. Troponin C binds to calcium ions and regulates the process of thin-filament activation during muscle contraction. Since there is no difference in troponin C between cardiac and skeletal muscle, it is not used as a cardiac marker. cTnT and cTnI are different molecules with different roles wherein, cTnT binds the troponin components to tropomyosin, whereas cTnI inhibits the interaction of myosin with actin. Nevertheless, elevation of either specifies myocardial damage. Presence of cTnT (>0.01 ng/mL) in blood indicates myocardial infarction (MI) (Wu et al., 1999). Ischemia is the most common cause of cardiac muscle damage, which triggers troponin release from the cytosolic and muscular pool. Troponin levels are elevated at 3-4 h after onset of myocardial injury, with peak values at about 24 h, after which it gradually declines. Troponins are also elevated in conditions other than MI viz., myocarditis, pericarditis, stroke, renal failure, pulmonary embolism, cardiac surgery, septicaemia, chemotherapy etc. (Sharma et al., 2004). One standardized assay exists for troponin T, while multiple assays are available for troponin I; each has a different cutoff value, as these assays target distinct epitopes.

DIABETIC MARKERS

Anti-Insulin Antibody

Anti-insulin antibody assay is carried out in blood to check whether the system has produced antibodies against insulin. These antibodies are described under two conditions viz., in insulin-naive individuals, wherein antibodies develop against endogenous insulin, which is an autoimmune phenomenon and in conditions where antibodies appear in insulin-treated patients. Earlier, insulin antibodies were detected in >95% patients treated with bovine and porcine insulin. This condition has greatly reduced due to the advent of human insulin and analogues. Antibodies produced against insulin are usually polyclonal IgG. In normal conditions, insulin antibodies are not detected in blood. Presence of IgG and IgM antibodies against insulin makes insulin less effective, which may lead to insulin resistance. High levels of IgE antibody against insulin indicate the allergic response that has developed due to insulin administration and this may lead to skin reactions (Greenfield et al., 2009).

C- Peptide

C-peptide assay is an index of pancreatic beta cell function. C- peptide or connecting peptide has 31 amino acids. During the course of insulin biosynthesis, proinsulin is cleaved into insulin and C-peptide. Both are stored in the secretory granules of the pancreatic β cells and eventually released together in equimolar amounts. C-peptide plays an important role in facilitating the correct folding of insulin and formation of its disulfide bridges. The main use of C-peptide is to evaluate hypoglycaemia. It is a valuable tool in elucidating the pathophysiology of type 1 and type 2 diabetes. The biological half life of C-peptide is ~30 min in healthy individuals and longer in subjects with type 2 diabetes. Unlike insulin, C-peptide escapes hepatic retention and it is eventually catabolised primarily by renal cortex, with a small fraction excreted in urine. Higher levels are observed in overweight individuals (Yosten et al., 2014). High levels of both C- peptide and blood glucose correlate with type 2 diabetes or insulin resistance whereas low levels of C-peptide with high blood glucose indicate type 1 diabetes.

HbA1c

HbA1c (glycated haemoglobin) is now formally endorsed in many countries as a diagnostic tool for diabetes mellitus (Florkowski, 2013). Hemoglobin found in red blood corpuscles (RBCs) binds to glucose present in the blood and forms glycated hemoglobin. Since the average life span of RBCs is 120 days, measurement of HbA1c reflects the mean glucose concentration of the past 3 months. It is ideal for diabetic patients to maintain their levels within 7%, to avoid other complications like cardiovascular disorders, kidney disorders etc.

Insulin

Insulin is produced by the beta cells of the pancreatic islets. This peptide hormone's important function is to transport glucose, the body's main source of energy, from the blood to within cells. It also stimulates the production and storage of triglycerides and proteins. Being an important blood glucose regulator, it is released from the pancreas after a high glucose meal whereas a low blood glucose level prevents its release. Production of low levels of insulin or insulin resistance may lead to cell starvation, while high levels of insulin lead to hypoglycemic conditions. Insulin level is measured in blood to diagnose low blood sugar (hypoglycemia). Normal values of insulin should be <25 mIU/L. High insulin level under normal blood sugar represents a hard working pancreas due to insulin resistance syndrome, which is common among people with obesity, polycystic ovary syndrome etc. Higher values of insulin correlate with insulinoma, injection of high dose of insulin, insulin resistance (which appears in type 2 diabetes, metabolic syndrome) etc, while lower values indicate type 1 or type 2 diabetes, hypopituitarism etc. Usually, insulin levels are diagnosed along with C-peptide, blood sugar, anti-insulin antibody tests etc.

ENDOCRINOLOGY MARKERS

Glycoprotein Hormone (GPH) Family

GPH family consists of three gonadotropins viz., luteinizing hormone, follicle-stimulating hormone and human chorionic gonadotropin and a fourth non-gonadotropin member, thyroid-stimulating hormone (Jiang et al., 2014).

Luteinizing Hormone (LH) The gonadotrophic cells of the anterior pituitary produce LH, which plays a major role in the regulation of ovulation in women. LH is a heterodimeric glycoprotein comprising α and β subunits. During a normal menstrual cycle, a surge in LH spurs ovulation from the dominant follicle. This process gets initiated during the mid-follicular phase, when LH stimulates androgen production in thecal cells. The LH surge results in ovulation and during post-ovulation, LH promotes progesterone production and this supports the development of corpus luteum. In men, LH stimulates testosterone production by Leydig cells of the testis, which plays a role in sperm production. Detection of LH levels in blood helps to diagnose the cause of infertility in men and women. In women, the levels of LH vary during the menstrual cycle with higher values at mid cycle peak. Abnormal increased levels in women may be due to non-ovulation, polycystic ovary syndrome, during or after menopause, Turner syndrome etc., while in men, it could be due to absence or non-functioning of testes, Klinefelter syndrome, overactive endocrine glands etc. Higher levels in children are diagnosed during precocious puberty. Lower values in both the sexes could be due to stress, low body weight, pituitary gland failure etc. (Pagana and Pagana, 2010; Choi and Smitz, 2014).

Follicle Stimulating Hormone (FSH) FSH is secreted by the anterior pituitary gland in response to the gonadotropin-releasing hormone, produced by the hypothalamus. FSH is a glycoprotein-gonadotropin and it helps to control the menstrual cycle and production of eggs by the ovaries. The level of FSH varies through the menstrual cycle and it is slightly elevated before ovulation. In men, FSH controls the production of sperms and its levels remain constant. Determination of FSH levels helps to diagnose the functioning of the sex organs in male and female and it helps to find out the cause of infertility. Normal FSH level will differ depending on a person's age and gender.

High FSH levels in women may be due to menopause, tumor in pituitary gland, Turner's syndrome etc, while low levels indicate underweight, non-ovulation, pregnancy, problems in pituitary gland etc. High levels in men may be due to Klinefelter syndrome, non-functioning of testicles, damage to testicles due to alcohol, chemotherapy etc. Low levels indicate deficiency in the production of the hormone by pituitary or hypothalamus. High FSH levels in boys or girls may mean that puberty is about to start. (Pagana and Pagana, 2010; Gruber and Farag, 2011). Levels of FSH in different conditions are shown in Table 3.2.

Human Chorionic Gonadotropin (hCG) hCG is a heterodimeric glycoprotein consisting of α and β subunits, produced by the placenta during pregnancy. Its α subunit is identical to that of LH, FSH and thyroid stimulating hormone (TSH). hCG can be detected both in blood and urine during pregnancy. hCG levels vary during pregnancy and it also affects the development of fetus. hCG levels increase during early pregnancy, peaks around 14-16th week after which there is a gradual decrease. hCG test is usually done to check pregnancy and sometimes as a screening test for birth defect (Cole 2009; Fischbach and Dunning, 2009). hCG levels in different conditions could be seen in Table 3.2.

Thyroid Stimulating Hormone (TSH) TSH is a glycoprotein hormone, produced by the thyrotrope cells of the pituitary gland and it regulates the endocrine function of the thyroid. TSH stimulates thyroid gland to produce thyroxine (T_4) and triiodothyronine (T_3). TSH level is used for evaluating thyroid function and/or disorders viz., hypo and hyperthyroidism. High level of TSH correlates with hypothyroidism, underactive thyroid gland, problems in pituitary gland etc. Low level associates with hyperthyroidism, pituitary gland damage etc. Nevertheless, abnormal TSH values have to be always correlated with free T_4 and T_3 levels to have a clear picture. Also, the values are altered during pregnancy, liver disease, systemic illness etc. (Rugge et al., 2015; Springer et al., 2017).

Parathyroid Hormone (PTH) PTH is produced by the parathyroid glands, which lie behind the thyroid glands. It is released in response to low calcium levels. PTH causes the bones to release more calcium into the blood and reduces the amount of calcium released by the kidneys into the urine. Also, vitamin D is converted to a more active form, causing the intestines to absorb more calcium and phosphorus. PTH is generally assayed along with calcium to identify hyperparathyroidism, to find out the cause for abnormal blood calcium levels and to evaluate parathyroid function. Low levels of PTH may be caused by damage to the parathyroid gland, high levels of vitamin D or calcium etc. High levels indicate hyperparathyroidism, which could be due to a benign parathyroid tumor, low levels of calcium etc *(Poole and Reeve, 2005)*.

Growth Hormone (GH) GH is a peptide hormone produced by the pituitary gland which is essential for the normal growth and development in children. It plays an important role in energy and food metabolism, production of RBCs, muscle mass etc. GH assay is done in children and adults having abnormal growth. High level of GH in childhood can cause a child to grow taller than normal (gigantism) while its deficiency causes a child to grow less than normal (dwarfism). Both conditions can be treated if found early. In adults, higher level of GH is caused by a noncancerous tumor of the pituitary gland (adenoma) triggering the abnormal growth of the bones of the face, jaw, hands, and feet (acromegaly). The levels of GH vary during day and night times; hence, single measurement of GH level in blood is difficult to interpret and hence, multiple samples are drawn.

Insulin-like Growth Factor – 1 (IGF-1) IGF-1 is a hormone that, along with GH, helps promote normal bone and tissue growth and development. IGF-1 is produced by the liver and skeletal muscle as well as many other tissues in response to GH stimulation. However, unlike GH, the level of

IGF-1 is stable in the blood throughout the day. This makes IGF-1 a useful indicator of average GH levels and the IGF-1 test is often used to evaluate GH deficiency or GH excess. GH/IGF-1 are master regulators of cellular function and impaired GH and IGF-1 signaling in aging/disease states leads to significant alterations in tissue structure and function, especially in the brain (Ashpole et al., 2015). Decreased levels of GH and IGF-1 are associated with slow growth and delayed child development while increased levels are associated with gigantism or acromegaly.

Prolactin Prolactin is a hormone produced by the anterior pituitary gland and it plays a primary role in lactation (breast milk production). Hence, it is generally present in low amounts in men and non-pregnant women; but, high in pregnant and breast feeding women. During pregnancy, the hormones prolactin, estrogen and progesterone stimulate breast milk development. Following childbirth, prolactin helps to initiate and maintain the breast milk supply. In women, who do not breastfeed, the prolactin level soon drops back to pre-pregnancy levels. In breast feeding mothers, suckling by the infant plays an important role in the release of prolactin. There is a feedback mechanism between how often the baby nurses and the amount of prolactin secreted by the pituitary as well as the amount of milk produced. Assay for prolactin is done in case of galactorrhea (breast milk production without pregnancy), absence of menstrual periods in women and to diagnose the cause of erectile dysfunction in men. The levels of prolactin vary over a 24 h period, rising during sleep and peaking in the morning. Ideally, a person's blood sample should be drawn 3 to 4 h after wake up. Higher levels may indicate prolactinoma (pituitary gland tumor), kidney disease, hypothyroidism etc., whereas lower levels could be due to hypopituitarism (Bole-Feysot et al., 1998).

Gastrin Gastrin was the first gastrointestinal hormone to be measured in plasma. It is produced by the G cells, in the stomach lining. When food is taken, the stomach becomes less acidic, stimulating the release of gastrin. Gastrin, in turn, stimulates parietal cells to produce gastric acid. As acidity increases in the stomach, food is broken down and gastrin release is suppressed. This feedback system normally results in low to moderate concentrations of gastrin in the blood. Gastrin levels are measured in the blood to evaluate an individual with recurrent peptic ulcer and/or other serious abdominal symptoms (Rehfeld et al., 2012).

Renin   Renin is a central hormone in the control of blood pressure and various other physiological functions. The juxtaglomerular cells (JGCs) of the kidney constitute the most important source of circulating renin (Persson 2003). Renin, along with aldosterone, plays an important role in maintaining normal sodium and potassium levels in the blood and thereby regulates blood pressure. Renin is assayed along with aldosterone to find out the cause of high blood pressure. The values of renin vary with the diet (low sodium or normal sodium diet).

Sex Hormone Binding Globulin (SHBG)   SHBG is a glycoprotein produced by the liver and it binds to the sex steroids viz., estrogen, testosterone and dihydrotestosterone in circulation. It is involved in the transport of sex steroids in blood and its concentration is a major factor regulating their distribution between the protein-bound and free states. Changes in SHBG levels can affect the amount of hormone that is available to be used by the body's tissues. SHBG secretion is suppressed by insulin and hence, a decrease in the levels of this protein is observed in states of insulin resistance. Circulating low level of SHBG is a potential marker of type 2 diabetes mellitus (Le et al., 2012) and gestational diabetes mellitus (Tawfeek et al., 2017). High plasma SHBG concentrations are observed in a number of diseases viz., hyperthyroidism, hypogonadism, androgen insensitivity and hepatic cirrhosis in men. Low concentrations are seen in myxoedema, hyperprolactinaemia and syndromes of excessive androgen activity. SHBG concentrations are also affected by androgens, estrogens, thyroid hormones and anti-convulsants. Measurement of SHBG is useful in the evaluation of mild disorders of androgen metabolism and it enables identification of those women with hirsutism who are likely to respond to estrogen therapy. Testosterone:SHBG ratios correlate well with both measured and calculated values of free testosterone and help to discriminate subjects with excessive androgen activity from normal individuals. Normal values are shown in Table 3.2.

GENERAL BIOCHEMICAL MARKERS

Adiponectin

Adipose tissue is an endocrine organ that produces biologically active molecules known as "adipocytokines". They are protein hormones with pleiotropic functions involved in the regulation of energy metabolism as well as in appetite, insulin sensitivity, inflammation, atherosclerosis, cell proliferation etc. Adiponectin and leptin are the major adipocytokines among which adiponectin is widely recognized for its antidiabetic, anti-inflammatory, anti-atherogenic and cardioprotective effects. Adiponectin modulates a number of metabolic processes including glucose regulation and fatty acid oxidation. Adiponectin circulates at high concentrations (5 – 30 μg/ml) in blood accounting for 0.01% of total serum proteins. Low levels of adiponectin correlate with obesity, metabolic syndrome, insulin resistance and type 2 diabetes (Lihn et al., 2005; Nigro, 2014).

Leptin

Leptin is an adipokine secreted by the adipose tissue, which helps to regulate energy balance by inhibiting hunger. Hence, it's called as the "satiety hormone". Leptin also plays a significant role in the regulation of lipid and carbohydrate metabolism. Leptin is found to decrease appetite, increase energy expenditure, suppresses insulin synthesis and secretion and increases insulin sensitivity (Huerta, 2006). Leptin regulates appetite by signaling hunger satisfaction to receptors in the hypothalamus of the brain. Low level of leptin triggers hunger and increases food consumption while an increased level diminishes hunger. Hence, the levels of leptin follow a circadian rhythm in our body. The inherited disorder of leptin deficiency can cause severe obesity due to constant hunger and food consumption. Elevated leptin levels are associated with obesity in leptin resistance individuals. Women normally have higher leptin levels than men; especially, the concentrations are increased during pregnancy. The levels are shown in Table 3.2.

Albumin

Albumin is one of the extensively studied plasma proteins and it is widely used as a diagnostic tool in clinical conditions. Albumin is produced by the liver and it is normally present at a high concentration in blood plasma (3.5-5.5 g/dL) with a circulating half-life of about approximately 20 days. It is elevated mainly during acute dehydration viz., vomiting, diarrhoea, burns etc and it gets decreased in chronic liver diseases, nephrotic syndrome, intestinal malabsorption syndromes and protein calorie malnutrition (Busher, 1990). Yang et al., (2012) have further demonstrated that glycated albumin could also be used as a potential tool for diagnosis of diabetes mellitus.

Microalbumin

A small amount of albumin in the urine is referred to as microalbumin and the condition is known as microalbuminuria. Virtually, no albumin is excreted in the urine during the normal functioning of the kidneys. Presence of albumin in urine indicates damaged or diseased kidneys, which have lost their ability to conserve albumin and other proteins. This is frequently noticed in chronic diseases such as diabetes and hypertension with increasing amounts of protein in urine. Since albumin is one of the early proteins to be detected in the urine, measurement of microalbumin could be an early indicator of kidney disease. Usually, albumin and creatinine levels are measured in urine to obtain an albumin: creatinine ratio. Apart from kidney disease, urinary tract infection, vigorous exercise etc also can lead to positive test in urine.

Globulin

Globulin is made up of different proteins known as alpha, beta and gamma types. Some globulins are produced by the liver, whereas others are produced by the immune system. Globulin level is computed from the values of albumin and total protein [total protein – albumin (A) = globulin (G)]. High levels of globulin indicate blood diseases (viz., leukemia, hemolytic anemia, macroglobulinemia etc), autoimmune disease, kidney and liver disorders etc (Pagana and Pagana, 2010). Low levels of globulin may be caused due to malnutrition, nephrotic syndrome (protein loss by kidney), congenital immune deficiency etc. A high ratio of A/G signifies underproduction of immunoglubulins as noticed in leukemia. Whereas a low A/G ratio suggests overproduction of globulins, as seen in multiple myeloma or autoimmune diseases, or underproduction of albumin, such as may occur with cirrhosis, or selective loss of albumin from the circulation, as observed in nephrotic syndrome.

Total Protein

Serum total protein assay measures the total amount of protein in the blood expressed as g/dL. Low levels of total protein are associated with liver or kidney disorder, problems associated with protein absorption and digestion, malnutrition, celiac disease or inflammatory bowel disease. Higher level of total protein indicates dehydration, multiple myeloma, chronic inflammation, bone marrow disorders etc. Nevertheless, the abnormal levels of total proteins have to be further clarified with specific diagnosis (Pagana and Pagana, 2010).

Ferritin

Ferritin is a ubiquitous and conserved iron storage protein that plays a central role in iron metabolism and has the dual function of storing iron in bioavailable and non-toxic forms. Ferritin is composed of a 24-subunit protein cage with a hollow interior cavity. Apart from liver, spleen, skeletal muscles and bone marrow, ferritin is also found in extracellular fluids such as serum, synovial fluids and milk. Serum ferritin serves as a critical marker to detect total body iron status (Orino and Watanabe, 2008). High values can be due to hemochromatosis, Hodgkin's disease, leukemia, infection, inflammatory conditions etc. (Fan et al., 2013). Low values reflect iron deficiency which can be caused due to heavy blood loss due to menstrual bleeding, bleeding in intestinal tract and pregnancy. In rare cases, iron may be lost through the skin (psoriasis) or in urine. Abril – ulloa (2014) reports that increased ferritin levels are associated with the presence of metabolic syndrome.

HEMATOLOGY RELATED MARKERS

Hemoglobin (Hb)

Hemoglobin is an iron containing protein present in the red blood cells (RBCs) with its characteristic red colour. Hb transports oxygen from lungs to the body's tissue and returns carbon dioxide from the tissues back to the lungs. Hb is made up of four polypeptide molecules (globulin chains) that are connected together. Normal adult Hb consists of two alpha-globulin chains and two beta-globulin chains. In fetuses and infants, beta chains are not common and the Hb molecule is made up of two alpha chains and two gamma chains. As the infant grows, the gamma chains are gradually replaced by beta chains, forming the adult Hb molecule. Each globulin chain contains an important iron-containing porphyrin component termed "heme". Embedded within the "heme" is an iron atom that is vital in the transport of oxygen and carbon dioxide. Normal ranges of Hb vary depending upon the age and gender (Table 3.2). Low levels of Hb indicate anemia which could be due to excessive blood loss (trauma, chronic bleeding etc), nutritional deficiency (iron, folate or vitamin B_{12}), damage to bone marrow, kidney failure, thalassemia, hemolytic anemia etc. High levels of Hb indicate polycythemia which could be a result of lung disease, congenital heart disease, kidney tumors, dehydration, living at high altitudes etc.

Erythropoietin (EPO)

EPO is a glycoprotein hormone that regulates erythropoiesis and it is produced primarily in the kidneys. It plays a key role in the production of RBCs which carry oxygen in blood. EPO is produced and released by kidneys in response to low blood oxygen levels and it stimulates the production of RBCs in bone marrow. EPO remains active for a short period of time and is then eliminated in the urine. EPO is usually diagnosed in conditions viz., anemia, bone marrow disorder etc. Increased levels of EPO may lead to overproduction of RBCs (polycythemia or erythrocytosis), which can occur due to a kidney tumor or other cancers. Overproduction of RBCs can lead to increase in the thickness of blood, blood pressure, blood clots etc. Low levels of EPO can be correlated with damaged kidneys which may lead to anemia. EPO levels are usually assessed as a follow up to abnormal findings in complete blood counts viz., low hemoglobin, hematocrit and RBCs (Donnelly, 2001). Normal values are shown in Table 3.2.

Fibrinogen

Fibrinogen is a glycoprotein, a coagulation factor, essential for blood clot formation. It is produced in the liver and released into circulation along with several other coagulation factors. When tissue or blood vessels are damaged, the coagulation cascade is initiated by platelets, and clotting factors are activated to the site as needed, one after another. At the end of the cascade, fibrinogen is converted to fibrin by the action of thrombin. Fibrin is an insoluble protein that forms a threaded mesh over the injury site. Fibrinogen levels are measured to assess bleeding disorders, cardiovascular disorders (CVD) etc. Two types of tests are available to evaluate fibrinogen: a fibrinogen activity assay evaluates the function of fibrinogen to form a blood clot while a fibrinogen antigen assay measures the amount of fibrinogen in blood. Decreased levels are associated with chronic liver disease, afibrinogenemia, hypofibrinogenemia etc. Since it is an acute phase reactant, increased levels are seen during inflammation, trauma, infection, coronary heart disease, peripheral arterial disease etc. (Monroe et al., 2010).

Factor VIII (FVIII)

FVIII, an essential blood coagulation protein, is a key component of the fluid phase blood coagulation system. It is also called as coagulation factor VIII or anti-hemophilic factor A. This glycoprotein is synthesized mainly in hepatocytes, and also in kidneys, endothelial cells and lymphatic tissue. It is one of the largest coagulation factors present in blood in association with von Willebrand factor (vWF) in a non-covalent complex. vWF prevents factor VIII from premature proteolysis and transfers it to sites of endothelial injury. It has a half life of about 8 – 12 h (Pisarek et al., 2016). Factor VIII, an acute phase reactant, is assayed, when there is a suspicion of haemophilia.

von Willebrand Factor (vWF)

vWF is a protein and it is one of the components of the coagulation system that work together, and in sequence, to stop bleeding within the body. vWF affects the clotting process by influencing the availability of factor VIII. vWF carries factor VIII and releases it at necessary sites. vWF test is done to determine the cause of excessive and repeated episodes of bleeding. Deficiency of vWF is found in von Willebrand disease (VWD), which is a common inherited bleeding disorder (Ragni et al., 2016).

Since vWF is an acute phase reactant, its levels get temporarily increased with infections, inflammation, trauma, and with physical and emotional stressors. It is also increased during pregnancy. Two types of tests are used to assess vWF.

- vWF antigen – this test measures the amount of the vWF protein present in the blood.

- vWF activity (also called Ristocetin Cofactor) – this test determines whether the protein is functioning properly.

Haptoglobin

Haptoglobin is a plasma protein primarily produced by the liver which binds to free hemoglobin (Hb) released by the lysing RBCs. Released Hb is toxic due to the redox activity of its heme center. Haptoglobin chaperones Hb subunits to the macrophages for safe degradation. Since haptoglobin levels become depleted in the presence of large amounts of free hemoglobin, decreased haptoglobin is a marker of hemolysis (Shih et al., 2014). Miyoshi and Nakano (2008) have reported that fucosylated haptoglobin is a novel marker for pancreatic cancer.

Hemopexin

Hemopexin is a predominant heme binding protein present in the blood. It also prevents heme toxicity, the major impetus responsible for the pathology in sepsis, sickle cell disease and other hemolytic conditions. Hemopexin plays a haptoglobin independent role in genetic and non-genetic hemolytic diseases. In plasma, hemopexin targets heme to the liver parenchymal cells for heme catabolism, iron storage and re-distribution. In barrier tissues, hemopexin targets heme to brain neurons (Hahl et al., 2013). A decreased level of hemopexin is noticed in case of intravascular inflammation and extravascular hemolysis such as that generated by extracorporeal circulation during cardiopulmonary bypass (CPB) and blood transfusions (Smith and McCulloh, 2015).

HEPATITIS MARKERS

Anti-HAV

Hepatitis A virus (HAV) is the causative agent of contagious liver infection, which is characterized by inflammation and enlargement of the liver. Anti-HAV blood test detects the presence of antibodies in blood that are directed against HAV. There are two different classes of hepatitis A antibodies, IgM and IgG. When a person is exposed to HAV, the body first produces hepatitis A IgM antibodies. These antibodies typically develop 2 to 3 weeks after first being infected (and are detectable before the onset of symptoms) and persist for about 3 to 6 months. Hepatitis A IgG antibodies are produced within 1 to 2 weeks of the IgM antibodies and usually persist for life. Because hepatitis A IgM antibodies develop early in the course of infection, a hepatitis A IgM assay is usually considered positive for a current or recent infection of hepatitis A. A HAV IgG test may be used to determine if a person has been infected in the past and has some immunity to the disease. A total hepatitis A antibody test detects the presence of both the IgM and IgG antibodies and thus, can identify current and past infections (Lee et al., 2013).

Anti- HCV

Hepatits C virus (HCV) causes an infection of the liver, characterized by liver inflammation and damage. Detection of HCV can be done by checking for antibodies in blood that are produced against HCV infection, by detecting the presence of viral RNA and also by HCV genotyping. Presence of HCV antibodies indicates a current infection or past exposure to HCV (Hyun et al., 2016).

HDAg and Anti-HDV

Serological markers used to ascertain hepatitis D virus (HDV) infection are the HDV antigen (HDAg) and the HDV antibody (anti-HDV). IgM anti-HDV is detectable during the 'window' phase of the infection, before the development of IgG anti-HDV; when detected at high titers, it is indicative of chronic infection. The development of anti-HDV antibodies, both of IgM or IgG subtype, is universal in individuals with HDV infection and it is an expression of the innate and adaptive response by the infected host.

Anti-HEV

Hepatitis E virus (HEV) usually causes an acute, self-limited infection. HEV-infected patients develop symptoms of hepatitis with appearance of anti-HEV IgM antibody in serum, followed by detectable anti-HEV IgG within a few days. Anti-HEV IgM may remain detectable up to 6 months after onset of symptoms, while anti-HEV IgG usually persists for many years after infection. Anti-HEV IgM is the serological marker of choice for diagnosis of acute HEV infection. Positive results of anti-HEV confirm the presence of acute or recent (in the preceding 6 months) hepatitis E infection (Aggarwal and Jameel, 2011; Hoofnagle et al., 2012).

Hepatitis B Panel of Blood Tests

Hepatitis B Virus Surface Antigen (HBsAg):   Hepatitis B virus (HBV) is the major causative agent of chronic hepatitis, hepatic decompensation, liver cirrhosis, and hepatocellular carcinoma (HCC). HBsAg is an important serum marker used for the clinical diagnosis and prognosis of HBV infection (Chen et al., 2014).

Hepatitis B Virus Surface Antibody (HBsAb):   Presence of HbsAb (antibody against hepatitis B surface antigen) indicates that the person is protected against hepatitis B virus, which could be due to vaccination or due to a previous infection. HBsAb levels of ≥ 10 IU/L indicate protective immunity (Norouzirad et al., 2014).

Hepatitis B Virus Core Antibody (HBcAb): Presence of HBcAb in blood indicates a present or past Hepatitis B infection. The core antibody does not provide any protection against the hepatitis B virus (unlike the surface antibody described above). This test can only be fully understood by knowing the results of the first two tests viz., HBsAg and HBsAb (Al-Mekhaizeem et al., 2001).

HIV MARKERS

Human Immunodeficiency Virus (HIV) Antigen and Antibody

HIV is the cause of acquired immunodeficiency syndrome (AIDS). HIV screening tests detect the HIV antigen (p24) and/or HIV antibodies produced in response to an HIV infection in the blood. During the first few weeks of HIV infection, the amount of virus (viral load) and the p24 antigen level in the blood can be quite high. About 2-8 weeks after exposure to the virus, the immune system responds by producing antibodies directed against the virus that can be detected in the blood. As the initial infection resolves and the level of HIV antibody increases, both virus and p24 antigen levels decrease in the blood. By detecting both antibody and antigen, the combination test demonstrates that an infection is detected soon after exposure. These tests can detect HIV infections in most people by 2-6 weeks after exposure (Nasrullah et al., 2013).

CD4 and CD8

CD stands for "cluster of differentiation," a glycoprotein, present on the surface of T cells, which fights infection. CD4 cells are a type of lymphocytes (white blood cells), which form an important part of the immune system. They lead the attack against infection and they are also called as T4 cells or T-helper cells. CD4 and CD8 levels are assessed to measure the strength of the immune system, HIV infection and to monitor its treatment. CD4 cells are the major target for HIV. A CD4 count is typically reported as an absolute level or count of cells (expressed as cells per cubic millimeter of blood). A normal CD4 count ranges from 500–1,200 cells/mm^3 in adults and teens. Increase in viral load (HIV) leads to decrease in CD4 cells, while antiretroviral therapy helps to increase or stabilise the CD4 count. CD4 levels also decrease due to pneumonia, influenza, herpes simplex viral infection, cancer chemotherapy etc. CD4 counts are sometimes done in conjunction with CD8 counts. CD8 cells are another type of lymphocytes called as T-suppressor or cytotoxic T cells, whose primary role is to kill HIV infected cell and to block HIV replication. CD4/CD8 ratio is a reflection of immune system and a normal ratio is between 1 and 4. In HIV infections, CD4 cells are usually destroyed more rapidly than CD8 cells. Hence, the CD4/CD8 ratio decreases as HIV infection progresses, and the ratio should increase and/or stabilize when antiretroviral treatment is effective. Evaluation of CD4 and CD8 cells may also be used to help classify lymphomas (Zijenah et al., 2005; Ratnam et al., 2006).

LIPOPROTEIN MARKERS

Apolipoprotein

Apolipoproteins are proteins that bind to lipids to form lipoproteins, whose primary function is lipid transportation. They play an important role in maintaining the structural integrity and solubility of lipoproteins, in lipoprotein receptor recognition and the regulation of certain enzymes in lipoprotein metabolism. There are six major classes of apolipoproteins: A, B, C, D, E and H.

Apolipoprotein A (apo A): There are two forms of apo A viz., apo A-I and apo A-II. Apo A-I is found in greater proportion than apo A-II (about 3 to 1). Apo A-I is the main protein component of high density lipoprotein (HDL). HDL picks up the excess cholesterol from the tissues and transports it to the liver, where it is further recycled or excreted in bile. Thus, HDL helps the tissues to get rid of excess cholesterol and prevents fatty build up and plaque formation. Apo A activates the enzymes that load cholesterol from tissues into HDL and allows HDL to be recognized and bound by receptors in the liver at the end of the transport. The concentration of apo A-I can be measured directly and it tends to rise and fall with HDL levels. Low levels of apo A-I correlate with an increased risk of developing cardiovascular disorders (CVD). Apo A-I may also decrease due to chronic kidney disease, diabetes, obesity, smoking etc, while its increase may correlate with use of statins, pregnancy, weight reduction etc (Uehara and Saku, 2014).

Apolipoprotein B (apo B): Apo B is the main apolipoprotein of chylomicrons and low density lipoproteins (LDL). Increased level of apo B is related to CVD. Apolipoprotein B and the apo B/apo A1 ratios are thought to be better markers for assessing vascular disease risk and indicators of ischaemic stroke (Bhatia et al., 2006).

NEPHROLOGY MARKERS

The incidence of acute kidney injury (AKI) and chronic kidney disease (CKD) is increasing worldwide and it is associated with increased morbidity and mortality. Currently, kidney function is assessed with the measurement of urinary output, urinalysis, estimation of serum creatinine concentration (SCr) and blood urea nitrogen (BUN) level. In practice, the reliability of these biomarkers is suboptimal to detect kidney disease in early stages. Identifying patients early is the basic need to provide an early intervention to improve the outcome in both diseases. Kidney cells produce biomarkers upto 3 days before the clinical expression of AKI develops and this is an advantage to apply preventive measures and arrest the progression of the pathological process (Soni et al., 2009;Wasung et al., 2015). Important biomarkers for assessment of kidney function are: urinary/serum cystatin C, kidney injury molecule-1(KIM-1), neutrophil gelatinase associated lipocalin (NGAL) and fibroblast growth factor 23 (FGF 23).

Cystatin C

Cystatin C is a relatively small protein and is found in a variety of body fluids, including the blood. Cystatin C is filtered out of the blood by the glomeruli in the kidneys. A decline in kidney function leads to a decrease in the glomerular filtration rate (GFR) and an increase in cystatin C and other parameters of kidney function, such as creatinine and urea in the blood. The increases in these levels occur because the kidney is not able to properly filter the blood at a normal rate, causing their accumulation in the blood. On the other hand, improvement in kidney function leads to an increase in GFR, which would cause cystatin C, creatinine, and urea to decline as a result of the kidneys being able to effectively clear them from the blood. Because cystatin C levels fluctuate with changes in GFR, cystatin C test is used as one of the methods for evaluating kidney function (Wasung et al., 2015).

Kidney Injury Molecule 1 (KIM - 1)

Acute kidney injury (AKI) has been defined as a rapid decline in glomerular filtration rate. KIM-1 is a type I transmembrane glycoprotein. It is undetectable in healthy kidney tissue, but expressed at very high levels in proximal tubule epithelial cells in human kidneys after ischemic or toxic injury (Wasung et al., 2015).

Neutrophil Gelatinase Associated Lipocalin (NGAL)

Neutrophil gelatinase-associated lipocalin (NGAL) is a 25 kDa protein of the lipocalin family. It is synthesized in renal tubular, intestinal, hepatic, and pulmonary tissue. Its synthesis is upregulated markedly in tissue injury especially of the kidney. In proximal tubular injury as well as in distal tubular injury, an increased synthesis of NGAL is observed. In AKI, plasma NGAL levels rise, related to either concomitant hepatic, pulmonary, or intestinal tissue injury, coupled with decreased glomerular filtration of NGAL. NGAL is a good predictor of AKI development, severity, and therapeutic monitoring especially with postcardiac surgery, sepsis, renal replacement therapy, and rejection after kidney transplantation (Alhaddad et al., 2015).

Fibroblast Growth Factor 23 (FGF - 23)

Fibroblast growth factor 23 (FGF-23) is a bone-derived phosphaturic hormone produced primarily by bone osteocytes. FGF-23 regulates phosphorus and vitamin D metabolism by inhibiting phosphate reabsorption and 1,25-dihydroxy vitamin D (1,25-D) production by the kidney.

Measurement of FGF-23 aids in the diagnosis of patients with disorders associated with hypophosphatemia and hyperphosphatemia. Chronic kidney disease (CKD) is the most common condition associated with elevations of FGF-23. FGF-23 is also elevated in several inherited and acquired disorders of mineral metabolism viz., rickets and osteomalacia.

SPECIFIC PROTEIN MARKERS

Acute Phase Proteins

They are produced in the blood during pathological stimuli causing cell necrosis. They are mainly elevated during inflammation, infection and cancer. C-reactive protein (CRP), hs-CRP and serum amyloid A protein are the important acute phase proteins which serve as indicators of inflammation and tissue damage.

C-reactive Protein (CRP): CRP gets its name, due to its ability to bind to the Capsular-polysaccharide of the *Streptococci* and activates the complement system. CRP is produced and secreted by the liver as a soluble pentamer and it forms part of the innate immune rapid response to infectious agents. In healthy adults, its serum concentration is <10 mg/L; during acute inflammation with infection, its concentration increases very significantly (up to 100- to 1000-fold) (Trial et al., 2016). Wang et al., (2016) report that synovial fluid CRP is a good biomarker for the diagnosis of prosthetic joint infection with high sensitivity and specificity.

High Sensitivity C Reactive Protein (hs-CRP): Hs-CRP is an important cardiac marker, which is used to determine a person's risk for cardiovascular disorders. The hs-CRP assay detects lower levels of the protein than the standard CRP assay. It measures hs-CRP in the range from 0.5 to 10 mg/L. Higher levels of hs-CRP may predict the future risk of heart attack, stroke, sudden cardiac death and peripheral arterial disease, even when cholesterol levels are within an acceptable range.

Serum Amyloid A (SAA) Protein: SAA is an apolipoprotein, present in the serum of healthy adults at a concentration of < 10 mg/L (Targońska-Stępniak and Majdan, 2014). Its physiological function is the transport of liposoluble substances (notably oxidized cholesterol derivatives) to the liver. It is a positive acute phase protein (liver responds by producing it) that is synthesized under the regulation of inflammatory cytokines during both acute and chronic inflammation. It is present at a low concentration in normal state and may show increase upto 100 fold in many types of carcinogenesis and neoplastic diseases (Liu et al., 2012).

Alpha 1 Antitrypsin (A1AT)

A1AT is primarily produced by the hepatocytes of liver and it acts as a serine protease inhibitor with antiprotease and immunoregulatory activities. It plays a major role in protecting the lungs from the enzyme, neutrophil elastase which is released during a neutrophil mediated inflammatory process. Though the primary function of neutrophil elastase is against bacteria, it can also cause damage to lung tissue if not adequately balanced by A1AT. Low serum levels of this protein indicate alpha 1 antitrypsin deficiency (AATD), which is a hereditary disorder. In deficiency conditions, the liver produces abnormal shaped alpha 1 antitrypsin molecule, which gets stuck inside the liver leading to liver fibrosis and it is not transported to other parts of the body, where it is needed. Hence, the condition leads to liver disease, early-onset pulmonary emphysema, rare multi-organ vasculitis, necrotizing panniculitis and fibromyalgia (Santangelo et al., 2017). Normal serum levels of A1AT are 100-220 mg/dL by nephelometry and levels of < 57 mg/dL are usually observed in AATD with lung disease. An increase in levels could also be observed during acute inflammation, cancer and liver diseases in individuals without AATD, in pregnancy, in women on estrogen therapy and in persons receiving blood transfusions etc.(Stoller et al., 2017).

Antistreptolysin O (ASO)

ASO is an antibody targeted against streptolysin O, a toxic enzyme produced by Group A *Streptococcus* bacteria. "O" stands for oxygen labile, and this toxin causes hemolysis of RBCs. Group A streptococcus (GAS) is the cause of a wide range of acute suppurative diseases and following a latent period, non-suppurative diseases such as rheumatic fever and poststreptococcal glomerulonephritis. Diagnosis of the latter group requires evidence of preceding GAS infection. ASO and Anti- DNAse B are the most common of several antibodies that are produced by the body's immune system in response to a streptococcal infection with GAS. The blood test measures the amount of ASO in the blood. This test is done in individuals who have been infected with streptococcal infection by GAS, to help diagnose complications such as rheumatic fever or glomerulonephritis. A positive test of greater than 166 Todd units (or > 200 IU) could indicate recent or current group A, C, and G streptococcal infection (eg, upper airway infections, scarlet fever, toxic shock syndrome) and may support the diagnosis of post-streptococcal infection complications (eg, glomerulonephritis and rheumatic fever). The antibodies level starts to rise in 1-3 weeks after streptococcal infection, peaks in 3-5 weeks, and then goes back to insignificant level over 6-12 months. Rising titers over time are more indicative of infection than a single test and hence, repeating the test is recommended 10 days after the initial test (Bryant and Stevens, 2015; Low 2016). In order to optimise the diagnosis of preceding GAS infection, at least two sequential ASO measurements, together with simultaneous assay for Anti- DNAse B, a second antistreptococcal antibody, is recommended (Sen and Ramanan, 2014).

Anti-DeoxyribonucleaseB (Anti-DNAse B)

Anti- DNAse B is one of the most common of several antibodies that are produced by the body's immune system in response to a streptococcal infection with *GAS*. GAS (*Streptococcus pyogenes*) is responsible for causing streptococcal throat and a variety of other infections, including skin infections (pyoderma, impetigo, cellulitis). Usually, blood test for anti-DNAse B is done along with ASO test, to diagnose a previous GAS infection. A negative test is normal. However, some people may have a low concentration of this antibody viz., preschool children <60 units/mL, school age children <170 units/mL, and adults < 85 units/ml. A high level of antibody indicates a previous streptococcal infection and may be indicative of rheumatic fever or kidney problems (glomerulonephritis) (Bryant and Stevens, 2015).

Angiopoietin 2

Angiopoietin 2 is an angiogenic factor, which exerts a selective effect on endothelium by binding with Tie-2 receptor. Under physiological conditions, it is stored in Weibel-Palade bodies. Its release in the systemic circulation is triggered by factors viz., hypoxia, inflammation or mechanical injury. Concentrations of this factor reflect the extent of endothelial activation (El-Banawy et al., 2012). Sporek et al., (2016) have reported that Angiopoietin 2 serves as a predictor of acute pancreatic- renal syndrome in patients with acute pancreatitis.

Ceruloplasmin(Cp)

Cp contains 95% of serum copper and it is a protein of the α2-globulin fraction of human blood serum. The serum value of Cp is used as a diagnostic tool for neurodegenerative diseases (Vassiliev et al., 2005). Kelly et al., (2016) report that serum value of Cp less than 20 mg/dL is suggestive of Wilson's disease. Siotto et al., (2016) report that ceruloplasmin specific activity is a reliable marker of Cp status and it is helpful in the diagnosis of Alzheimer's disease.

Cryoglobulins

Cryoglobulins are circulating proteins, specifically immunoglobulins, that get precipitated below 37°C, dissolve when warmed and can cause multiorgan damage. There are three types of cryoglobulins: Type I (also called simple) is mostly associated with monoclonal gammopathy and/ or other hematological disorders. Type II and Type III (known as mixed cryoglobulins) are associated with infectious and systemic diseases (Motyckova and Murali, 2011). They are present in the range of 100-300 mg/L in normal individuals while high levels may indicate cryoglobulinemia, a signal that the body is producing abnormal proteins. Cryoglobulinemia can cause bruising, rashes, joint pain, tissue damage leading to skin ulcer and in severe cases, gangrene. Presence of cryoglobulins in blood could be associated with a variety of conditions viz., hepatitis C, AIDS, kidney disease, lyme disease, multiple myeloma, lymphoid leukemia, autoimmune diseases such as rheumatoid arthritis, systemic lupus erythematosus etc. (Ferri et al., 2002).

Complements

Complements is a system of several plasma proteins that migrate in the electrophoretic fraction of beta-globulins and it has a role in the defence against infections. The complement system is part of the body's innate immune system, which is non-specific and can quickly respond to foreign substance. Its principal role is to destroy foreign pathogens like bacteria and viruses. The proteins of complement system circulate in the blood in the form of inactive precursors, and when activated, they acquire enzymatic activities (mainly esterases) that allow them to damage the lipid bilayer of the (bacterial) cell membrane. Excessive response of the complement may also damage the cells of the organism. Complement is activated by the antigen-antibody complexes, C-reactive protein and some fungi, bacteria and viruses. An increased level of complements is observed during acute or chronic inflammation and low levels are observed during microbial infections. There is also an alternative activation pathway that does not involve immunocomplexes and it requires instead the specific serum protein *Properdin*. Several genetic defects of complements are known that cause increased propensity to infectious diseases. Normal values are presented in Table 3.2.

Immunoglobulins (Ig)

The major types of immunoglobulins present in our body are IgA, IgG, IgM, IgD and IgE. IgG is the most common type, with four different sub - classes that help to prevent infections. In normal serum, about 80% is IgG, 15% is IgA, 5% is IgM, 0.2% is IgD and a trace is IgE. Serum immunoglobulin levels (IgG, IgA, IgM) are elevated in conditions associated with chronic inflammation viz., rheumatoid arthritis, systemic lupus erythematosus etc and chronic infections viz., hepatitis C, HIV etc. IgM levels are increased in congenital or neonatal infection in newborn (Katzmann and Kyle, 2006). Monoclonal increase in one class of Ig (IgG or IgA or IgM) with or without decrease in other two classes is correlated with multiple myeloma, chronic lymphocytic leukemia, lymphoma etc.

Decreased Ig levels could occur due to (i) inherited disorders in which the body is not able to produce one or more classes of Ig (ii) conditions that affect the body's ability to produce Ig or increase the loss of Ig from the body. Nephrotic syndrome, renal diseases, severe burns, sepsis etc are the conditions that cause an abnormal loss or increased catabolism of Ig, while conditions which affect Ig production include malnutrition, alcoholism, immunosuppressive drugs, chemotherapy agents, malignancies, rheumatological diseases etc. (Jaffe et al., 2011).

Ig levels help in the diagnosis of liver diseases. Elevated IgM level is both a sensitive and specific marker for primary biliary cirrhosis. Increased IgG correlates with autoimmune hepatitis and elevated IgA points to alcoholic liver disease and liver cirrhosis during chronic hepatitis B infection (Lin et al., 2016).

Allergen specific IgE antibodies aid in the diagnosis of allergic diseases viz., rhinoconjunctivitis, atopic eczema, urticaria, anaphylaxis and food and drug allergies (Fall and Niessner, 2009). Normal values are shown in Table 3.2.

TUMOR MARKERS

Alpha Feto Protein (AFP)

AFP is a fetal specific glycoprotein normally produced primarily by the fetal liver. Normally, AFP levels decline rapidly, reaching undetectable levels (less than 10 ng/mL) within several months after birth (Ball et al., 1992). Blood test for AFP is usually done to diagnose and monitor therapy for cancer of the liver and testicles. More sensitive methods of detection have disclosed that this fetal protein persists in trace amounts throughout life and its level increases in maternal blood during pregnancy. It is notable that the synthesis of AFP is increased in liver cancer cells and that high levels of this protein are present in serum. Elevated values of AFP have also been detected in human subjects with undifferentiated tumours of the testis and ovary. A fall back to normal levels has been noted in cases of complete remission after surgery and a return to high levels is seen in patients developing metastases. In some patients with hepatitis, a temporary rise in the level of AFP has also been observed. Detection of high levels of AFP in amniotic fluid has proved to be of great value in the prenatal diagnosis of neural-tube defects. Low levels of AFP in pregnant women can also indicate chromosome abnormality such as Down's or Edwards syndrome in the developing fetus. Abnormal levels have also been found in the amniotic fluid or in maternal serum in cases of spontaneous abortion. Such measurements are now being assessed as a method of monitoring abnormal pregnant.

Beta 2 Microglobulin(β2M)

β2M is a small protein (11,800 Dalton), present in nearly all nucleated cells and most biological fluids, including serum, urine, and synovial fluids. Most nucleated cells in the human body carry MHC class I antigens that help the immune system identify self molecules. These antigens have a heavy chain and an associated light chain. β2M is a light chain protein which can be shed from the surface of nucleated cells into serum. Increased levels can be seen in a wide variety of disorders that involve increased cell turnover and/ or activation of the immune system. Increased level of this protein in blood is observed in acute and chronic inflammations, infections due to virus viz., cytomegalovirus and HIV, several malignancies (lymphoma, leukemia, multiple myeloma etc), amyloidosis associated with hemodialysis, Crohns disease etc. A low level of β2M is observed in renal tubular acidosis. In rare cases, cerebrospinal fluid β2M level is used to assess disease involved with the central nervous system. Abnormal values of β2M in urine indicate renal filtration and reabsorption disorders. Serum β2M levels during hospital discharge aids in predicting long term mortality and graft loss in kidney transplantation recipients (Li et al., 2016).

Bence Jones Protein (BJ Protein)

BJ protein is a light chain of immunoglobulins produced by the white blood corpuscles (WBC). Normal healthy individuals lack this protein in the urine; but, presence of this protein in urine indicates multiple myeloma. Presence of BJ protein may also be due to amyloidosis, chronic lymphocytic leukemia, white blood cell cancer (macroglobulinemia) etc (Brigden et al., 1990).

Cancer Antigen 125 (CA 125)

CA 125 is a heterogeneous cell membrane glycoprotein and it is related to malignant conditions such as ovarian, uterine, lung or pancreatic cancers (Cragun, 2011). CA 125 is used as a tumor marker since it is present on the surface of most, but not all ovarian cancer cells. CA 125 is usually measured in blood to monitor the treatment progress for ovarian cancer or to detect recurrence and it is not generally recommended for screening of ovarian cancer, due to false-positive results. Kim et al., (2015) report that preoperative CA 125 is a significant indicator of curative resection in gastric cancer patients.

Cancer Antigen 15-3 (CA 15-3)

CA 15-3, a glycoprotein, is a tumor marker, elevated in breast, ovarian, pancreatic, lung and colorectal cancers. Levels of CA 15-3 are assessed in blood to monitor the response to treatment of invasive breast cancer and to help watch for recurrence of the disease. Though CA 15-3 is expressed by normal breast cells, its production is increased in breast cancer, especially in case of metastatic breast cancers. But, CA 15-3 is not recommended as a screening test to detect breast cancer because of its non-specificity (Fejzic et al., 2015).

Cancer Antigen 19-9 (CA 19-9)

CA 19-9 is a tumor associated mucin glycoprotein antigen. Its level is assessed in blood to monitor the treatment response to pancreatic cancer and to check its recurrence. At times, it also aids in the diagnosis of pancreatic cancer. CA 19-9 is synthesized by normal human pancreatic and biliary ductal cells and by gastric, colon, endometrial and salivary epithelia. CA 19-9 is present in small amounts in serum, and can be over expressed in several benign gastrointestinal disorders. Importantly, it exhibits a dramatic increase in its plasma levels during neoplastic disease (Scara et al., 2015). In general, before surgery, the higher the CA19-9 level is, the larger the tumor is and the less chance that the tumor is resectable. For the purpose of evaluating treatment, a decreasing or stable CA19-9 level generally indicates an improved prognosis and an increasing level indicates the progression of disease. CA 19-9 is elevated in the blood of 50-75 % of pancreatic cancer patients. Jo et al., (2013) report that CA 19-9 is also elevated in patients with metastatic or recurrent gastric cancer.

Cancer Antigen 72-4 (CA 72-4)

CA 72-4 is a glycoprotein found on the surface of many tumor cells viz., ovary, breast, colon, lung and pancreas. This tumor marker is used to diagnose and monitor many neoplastic diseases. It is widely used for screening and to identify relapses of gastric and ovarian cancer (Goral et al., 2007).

Carcinoembryonic Antigen (CEA)

CEA is an oncofetal glycoprotein, which is involved in cell adhesion. CEA is normally found in the tissue of a developing baby in the womb. The blood level of this protein disappears or becomes very low after birth. In adults, an abnormal level of CEA may be a sign of cancer. Blood test measures the amount of CEA in the blood to evaluate individuals diagnosed with cancer. The normal range is 0 to 2.5 µg/L. In smokers, the normal range is 0 to 5 µg/L. Originally, it was thought that CEA was a specific marker for colon cancer; but, further study has shown that an increase in CEA may also be seen in a wide variety of other cancers. CEA can also be increased in some non-cancer-related conditions such as inflammation, cirrhosis, peptic ulcer, ulcerative colitis, rectal polyps, emphysema and benign breast disease, and in smokers. For this reason, it is not useful as a general cancer screening tool; but, it does have a role in evaluating response to cancer treatment. Increased levels of CEA indicate recurrence of cancer in treated patients or it could be due to cancer of the breast, colon, lung, pancreas, thyroid, reproductive and urinary tracts (Duffy 2001; Pagana and Pagana 2011). Kaplan et al., (2015) report that CEA levels may be a secondary marker for diagnosing bipolar disorder.

Prostate Specific Antigen (PSA)

PSA is a glycoprotein (mol.wt. 30,000 da), which belongs to the kallikrein family. PSA is produced by the glandular epithelium of the prostate and secreted in the seminal fluid. It is also present in urine and blood. PSA levels are elevated in men with prostate cancer. In addition, PSA levels also tend to increase in persons with prostatitis (inflammation of the prostate) and benign prostatic hyperplasia (BPH) (Kantoff and Talcot, 1994). PSA test would help to determine prostate cancer early, even in men who are symptomless. PSA level also enables the detection of metastases or the disease persistence after cancer therapy. The free-PSA test measures the percentage of unbound PSA while the PSA test measures the total of both free and bound PSA. The normal range of total PSA in males is between <2.0 ng/mL – 7.2 ng/mL (differs based on age). When total PSA is in the range of 4.0-10.0 ng/mL, a free:total PSA ratio < or = 0.10 indicates 49% to 65% risk of prostate cancer depending on age; a free:total PSA ratio >0.25 indicates a 9% to 16% risk of prostate cancer, depending on age. The total PSA range of 4.0 to 10.0 ng/mL is considered as a diagnostic "grey zone," in which the free:total PSA ratio helps to determine the relative risk of prostate cancer.

CD 20 Antigen

CD 20 is a B cell differentiation antigen widely expressed in B cell development from early pre-B until mature B cell stage, but lost on differentiation in to plasma cells. This selective expression on mature B cells but not on precursors such as stem cells or on antibody-secreting plasma cells makes it an attractive target. Measuring the amount of CD20 antigen on blood cells may help to diagnose cancer or plan the treatment (Fang et al., 2013).

Progesterone and Estrogen Receptors

Estrogen receptors (ER) and progesterone receptors (PR; also called PgR) are shown to over express in breast cancer cells (Yip and Rhodes, 2014). Cancer cells with these receptors depend on estrogen and related hormones, such as progesterone, to grow. Estrogen and progesterone influence many hormonal functions in women, such as breast development. If breast cancer cells have estrogen receptors, the cancer is called ER-positive breast cancer. If breast cancer cells have progesterone receptors, the cancer is called PR-positive breast cancer. If the cells do not have either of these two receptors, the cancer is called ER/PR-negative. About two-thirds of breast cancers are ER and/or PR positive. Testing for hormone receptors is important because the results help to decide whether the cancer is likely to respond to hormonal therapy or other treatments.

Human Epidermal Growth Factor Receptor 2 (HER2)

Human epidermal growth factor receptor 2 (HER2) is a member of the epidermal growth factor receptor family having tyrosine kinase activity. Dimerization of the receptor results in a variety of signaling pathways leading to cell proliferation and tumorigenesis. Amplification or overexpression of HER2 occurs in approximately 15–30% of breast cancers and 10–30% of gastric/gastroesophageal cancers and HER2 serves as a prognostic and predictive biomarker. HER2 overexpression has also been seen in other cancers like ovary, endometrium, bladder, lung, colon, and head and neck. Checking the amount of HER2 on some types of cancer cells may help plan treatment (Iqbal and Iqbal, 2014).

Thyroglobulin (Tg)

Tg is a glycoprotein produced by both normal and cancerous thyroid cells. Treatment of thyroid cancer involves total thyroidectomy followed by radioactive iodine and thyroid hormone therapy. Radioiodine is given to destroy the remnant thyroid cells that are left after surgery. Measurement of thyroglobulin is helpful in the differential diagnosis of hyperthyroidism and in monitoring the course of differentiated or metastatic thyroid cancer (Gonzalez et al., 2014).

Table 3.2   Levels of different protein markers in normal and disease states (IU/L – International units/ Litre) (mIU/L - milli-international units per liter) Values are given based on the reports available. Individual lab to lab variations may be present.

Protein	Normal range		Abnormal range	Associated Disease/disorder
AUTO-IMMUNE MARKERS				
Anti-nuclear antibodies		Absent	Present	Auto immune disorder eg. Systemic lupus erythematosus
Anti-Sperm antibodies		Absent	Present	infertility
Anti-phospholipid antibodies	IgG (ELISA)	< 12 U/mL 12-18 U/ mL Equivocal	>18 U/mL	Antiphospholipid syndrome (APS), lupus, rheumatoid arthritis, HIV or rubella infection, tumors of lung, colon etc
	IgM (ELISA)	< 12 U/mL 12-18 U/ mL Equivocal	>18 U/mL	
Anti-Beta-2 glycoprotein 1	IgG	0 – 20 SGU	High	Antiphospholipid syndrome, thrombocytopenia
	IgM	0 – 20 SGU		
Rheumatoid factor		<15 IU/mL	>15 IU/mL	Rheumatoid arthritis
Cyclic citrullinated peptide antibody		< 5 U/mL	> 5 U/mL	Rheumatoid arthritis
CARDIAC MARKERS				
BNP		<100 pg/mL	>400 pg/mL	Heart failure
NT pro BNP		<125 pg/mL	>450 pg/mL	Heart failure
Cardiac troponin T		< 0.01 ng/mL	>0.01 ng/mL	Myocardial infarction, cardiac injury etc

Protein	Normal range		Abnormal range	Associated Disease/disorder
Cardiac troponin I		< 0.3 ng/mL	> 0.3 ng/mL	Myocardial infarction, cardiac injury etc
DIABETIC MARKERS				
Anti -Insulin antibody		Absent	Present	Diabetes
C - peptide	Overnight fasted state	1.1 – 4.4 ng/mL	> 4 ng/mL	Type 2 diabetes, insulinoma, chronic kidney disease, sulfonyl urea intoxication
	Postprandial state	1.2 – 3.4 ng/mL	Low	Type 1 diabetes, hypoglycemia, liver disease, pancreatectomy
HbA1c	Normal	4.0 – 6.0 %	7.0 – 8.0 % Fair control	Chance of acquiring diabetes
	Good control	6.0 – 7.0 %	> 8.0 % Poor control	Diabetes
Insulin	Fasting	< 25 mIU/L	High	Insulinoma, type 2 diabetes
			Low	Type 1 diabetes
ENDOCRINOLOGY MARKERS				
Luteinizing hormone	Women: Before menopause	5 – 25 IU/L	High in women	Fertility issues
	After menopause	14 – 53 IU/L	Low in both sexes	Pituitary gland failure
	Men	1.8–8.6IU/L		

Protein	Normal range		Abnormal range	Associated Disease/disorder
FSH	**Men:** Before puberty	0 - 5.0 mIU/mL	High or low	Infertility
	During puberty	0.3-10.0 mIU/mL		
	Adult	1.5 – 12.4 mIU/mL		
	Female Before puberty	1.5 – 12.4 mIU/mL		
	During puberty	0.3 - 10.0 mIU/mL		
	Women who are still menstruating	4.7 - 21.5 mIU/mL		
	After menopause	25.8 - 134.8 mIU/mL		
HCG	Men and non pregnant women	< 5 IU/L	High (men and non pregnant women)	Germ cell tumor, stomach cancer
	Pregnant women, 1 week of gestation	5-50 IU/L	High (pregnancy)	Multiple pregnancy, down syndrome
	Pregnant women, 2 weeks of gestation	50-500 IU/L		
	Pregnant women, 3 weeks of gestation	100-10000 IU/L		

Protein	Normal range		Abnormal range	Associated Disease/disorder
	Pregnant women, 4 weeks of gestation	1080-30000 IU/L		
	Pregnant women, 6-8 weeks of gestation	3500-115000 IU/L	Low (during pregnancy)	Miscarriage or ectopic pregnancy
	Pregnant women, 12 weeks of gestation	12000-270000 IU/L	Low (later stage of pregnancy)	Preeclampsia risk
	Pregnant women, 13-16 weeks of gestation	Upto 200000 IU/L		
TSH		0.27-5.35 µIU/mL	High	Hypothyroidism
			Low	Hyperthyroidism
Parathyroid hormone		10-65 pg/mL	High	Hyperparathyroidism
			Low	Damage to parathyroid gland
Growth hormone	Men	> 5 ng/mL	High or low	Growth related problems
	women	> 10 ng/mL		
	children	> 20 ng/mL		
IGF-1	11-18 years (male)	55-232 ng/mL	Low	Delayed growth
	11-18 years (female)	65-242 ng/mL	High	Acromegaly or gigantism

Protein	Normal range		Abnormal range	Associated Disease/disorder
Prolactin	Non-pregnant women	4-23 ng/mL	High	Prolactinoma
	Pregnant women	34 – 386 ng/mL	Low	Hypopituitarism
	Lactating mother	100 ng/mL		
	Lactating mother (6 months postpartum)	50 ng/mL		
	children	3.2 – 20 ng/mL		
	Men	3-15 ng/mL		
Gastrin	Adults	< 100 pg/ mL	High	Zollinger-Ellison (ZE) syndrome, stomach cancer, kidney failure
	Children	10-125 pg/mL	Low	Hypothyroidism
Renin		0.2 – 3.3 ng/mL/h	High	Blood pressure, kidney disease, haemorrhage, cirrhosis
			Low	Kidney disorder, Conn's syndrome
Sex hormone binding globulin (SHBG)	Male	11 – 78 nmol/L (ELISA)	Low	Diabetes mellitus
	Female	11-137 nmol/L (ELISA)	High	androgen metabolism, hyperthyroidism, hepatic cirrhosis

Protein	Normal range		Abnormal range	Associated Disease/disorder
GENERAL BIOCHEMICAL MARKERS				
Adiponectin		2.7 – 25.0 µg/mL	Low	Type 2 diabetes, obesity etc
Leptin	Male (18-25 years)	2.05 – 5.63 ng/mL	High	obesity
	Female (18 - 25 years)	3.7 – 11.1 ng/mL		
Albumin		3.5 – 5.5 g/dL	High	dehydration
			Low	Liver disorder, nephrotic syndrome, malnutrition, intestinal malabsorption syndrome
Microalbumin (urine)		Absent	30 – 300 mg/ 24 h	Microalbuminuria, kidney damage
Globulin		2.6 -4.6 g/L	High	Kidney and liver disorder, blood diseases
			Low	Malnutrition, nephrotic syndrome
Total protein		6.0-8.3 g/dL	High	Dehydration, multiple myeloma, bone marrow disorder
			Low	Liver and kidney disorders, malnutrition, protein malabsorption

Protein	Normal range		Abnormal range	Associated Disease/disorder
Ferritin	men	18 – 270 ng/mL		
	women	18 – 160 ng/mL	High	Leukemia, infection, inflammation etc
	children	7 – 140 ng/mL	Low	Iron deficiency
	Babies (1-5 months)	50 – 200 ng/mL		
	New borns	25 – 200 ng/mL		
HEMATOLOGY RELATED MARKERS				
Hemoglobin	New borns	17-22 g/dL	High	Polycythemia
	One week of age	15-20 g/dL	Low	Anemia, kidney failure
	One month of age	11-15 g/dL		
	Children	11-13 g/dL		
	Adult males	14-18 g/dL		
	Adult women	12-16 g/dL		
Erythropoietin		4.3 - 29.0 mIU/mL	High	Kidney tumor and other cancers
Fibrinogen		150 – 400 mg/dL	High	Cardiovascular disorder etc
			Low	Chronic liver disease etc
Factor VIII		50 – 150 %	Low	Haemophilia A, von Willebrand disease etc
			High	Inflammatory state
vWF	antigen	50-200 IU/dL	Low	von Willebrand disease
			High	Infection, inflammation

Protein	Normal range		Abnormal range	Associated Disease/disorder
Haptoglobin		30-200 mg/dL	Low	Haemolytic anemia
Hemopexin		500 – 1000 µg/mL	Low	hemolysis
HEPATITIS MARKERS				
Anti - HAV	Ig G, Ig M		Present	Acute hepatitis
Anti - HCV			Present	HCV liver infection
HDAg and Anti - HDV			Present	HDV liver infection
Anti - HEV			Present	HEV liver infection
HBsAg			Present	HBV liver infection
HBsAb			>10 IU/L	Protective immunity
HBcAb			Present	HBV liver infection
HIV MARKERS				
HIV (antigen and antibody)			Present	HIV infection (AIDS)
CD4		500 – 1200 cells/mm3	Low	Viral infection (HIV), pneumonia
CD4/CD8 ratio		1-4	Low	HIV infection
LIPOPROTEIN MARKERS				
Apolipoprotein A	Men	120 mg/dL	Low	Cardiovascular disease, chronic kidney disease, diabetes
	Women	140 mg/dL	High	Use of statins, pregnancy, weight reduction

Protein	Normal range		Abnormal range	Associated Disease/disorder
Apolipoprotein B		40 -125 mg/dL	High	Cardiovascular disease
NEPHROLOGY MARKERS				
Cystatin C	Serum	0.56 – 1.25 mg/L	High	Impaired kidney function
	Urine	0.03 – 0.18 mg/L		
Kidney injury molecule 1		< 1 ng/mL	High 3-7 ng/mL	Kidney injury
NGAL	Adult	107 ng/mL	High	Kidney injury
	Pediatric	117.6 ng/mL		
FGF23		74 RU/mL	High	Chronic kidney disease
SPECIFIC PROTEIN MARKERS				
CRP		< 10 mg/L	High	Acute inflammation with infection
hs-CRP		<1 mg/L	High	Cardiovascular disease, stroke
			1-3 mg/L	Average risk of CVD
			>3 mg/L	High risk of CVD
Serum amyloid A protein		< 10 mg/L	High	Inflammatory disorder, carcinogenesis
Alpha 1 antitrypsin		100 – 220 mg/dL	Low (<57 mg/dL)	Lung disease
			High	Cancer and liver diseases
ASO		< 160 Todd units/mL (or 200 IU)	> 166 Todd units/mL (or 200 IU)	Streptococcal infection

Protein	Normal range		Abnormal range	Associated Disease/disorder
Anti-DNase B		Negative	High	Streptococcal infection
Angiopoietin 2		Serum cut off <0.001 (ELISA)	High	Pancreatic-renal syndrome
Ceruloplasmin		20-35 mg/dL	<20 mg/dL	Wilson's disease
Cryoglobulins		100 – 300 mg/L	High	Cryoglobulinemia
Complements	C3- men	88-252 mg/dL	Low	Microbial infection, autoimmune disease
	C3- women	88-206 mg/dL	High	Acute or chronic inflammation
	C4- men	12-72 mg/dL		
	C4- women	13-75 mg/dL		
Immunoglobulins	Ig A	70 - 400 mg/dL	High	Chronic infection and inflammation, autoimmune disorders
	Ig G	700 – 1600 mg/dL		
	Ig M	40 – 230 mg/dL	Low	Inherited disorders
	Ig E	1.31 – 165.3 IU/mL		
TUMOR MARKERS				
Alpha feto protein		<15 ng/mL	High	Cancer of liver, testis and ovary
Beta 2 microglobulin	Serum or plasma	0.8 – 2.2 mg/L	High	Multiple myeloma, leukemia, lymphoma, viral infection (HIV, cytomegalovirus etc)
	Urine	0-0.3 µg/mL		
Bence jones protein			Present	Multiple myeloma

Protein	Normal range		Abnormal range	Associated Disease/disorder
CA 125		0-35 units/mL	High	Ovarian cancer
CA 15-3		< 30 U/mL	High	Breast cancer
CA 19-9		< 37 U/mL	High	Pancreatic cancer
CA 72-4		< 7 U/mL	High	Gastric and ovarian cancer
CEA		0-2.5 µg/L	High	Colon cancer
PSA (total)	< 40 years	$\leq$ 2.0 ng/mL	High	Prostate cancer, benign prostatic hyperplasia
	40 -49 years	$\leq$ 2.5 ng/mL		
	50 -59 years	$\leq$ 3.5 ng/mL		
	60 -69 years	$\leq$ 4.5 ng/mL		
	70 -79 years	$\leq$ 6.5 ng/mL		
	$\geq$ 80 years	$\leq$ 7.2 ng/mL		
CD 20		Absent	Present	Cancer
Progesterone receptor		Absent	Present	Breast Cancer
Eestrogen receptor		Absent	Present	Breast Cancer
Human epidermal growth factor receptor 2 (HER 2)		Absent	Present	Breast Cancer Gastric Cancer
Thyroglobulin (Tg)		< 10 ng/mL	< 2 ng/mL after thyroidectomy	Thyroid cancer (Papillary and follicular cancers)

References

* Abril-Ulloa V, Flores-Mateo G, Solà-Alberich R, Manuel-y-Keenoy B, Arija V. (2014). Ferritin levels and risk of metabolic syndrome: meta-analysis of observational studies. *BMC Public Health.*14:483.

* Aggarwal R, Jameel S. (2011). Hepatitis E. *Hepatology.* 54:2218-2226.

* Alhaddad OM, Alsebaey A, Amer MO, Hany El-Said H, and Salman TAH. (2015). Neutrophil Gelatinase-Associated Lipocalin: A New Marker of Renal Function in C-Related End Stage Liver Disease. *Gastroenterology Research and Practice.* 2015: Article ID 815484, 6 pages.

* Al-Mekhaizeem KA, Miriello M, Sherker AH. (2001). The frequency and significance of isolated hepatitis B core antibody and the suggested management of patients. *CMAJ: Canadian Medical Association Journal.* 165:1063-1064.

* Ashpole NM, Sanders JE, Hodges EL, Yan H, Sonntag WE. (2015). Growth hormone, insulin-like growth factor-1 and the aging brain. *Exp Gerontol.* 68:76-81.

* Ball D, Rose E, Alpert E. (1992). Alpha-fetoprotein levels in normal adults. *Am J Med Sci.* 303:157-159.

* Bhatia M, Howard SC, Clark TG, Neale R, Qizilbash N, Murphy MF, Rothwell PM. (2006). Apolipoproteins as predictors of ischaemic stroke in patients with a previous transient ischaemic attack. *Cerebrovasc Dis.* 21:323-328.

* Bidot CJ, Jy W, Horstman LL, Ahn ER, Yaniz M, Ahn YS. (2006). Antiphospholipid antibodies (APLA) in immune thrombocytopenic purpura (ITP) and antiphospholipid syndrome (APS). *Am J Hematol.*81:391-396.

* Bole-Feysot C, Goffin V, Edery M, Binart N, Kelly PA. (1998). Prolactin (PRL) and its receptor: actions, signal transduction pathways and phenotypes observed in PRL receptor knockout mice. *Endocr Rev.* 19: 225-268.

• Targońska-Stępniak B, Majdan M. (2014). Serum Amyloid A as a Marker of Persistent Inflammation and an Indicator of Cardiovascular and Renal Involvement in Patients with Rheumatoid Arthritis. *Mediators of Inflammation* 2014: Article ID 793628, 7 pages.

• Brigden ML, Neal ED, McNeely MD, Hoag GN. (1990). The optimum urine collections for the detection and monitoring of Bence Jones proteinuria. *Am J Clin Pathol.* 93:689-693.

• Bryant AE, Stevens DL. (2015). *Streptococcus pyogenes.* In: *Mandell, Douglas, and Bennett's Principles and Practice of Infectious Diseases* (Bennett JE, Dolin R, Blaser MJ, eds.) 8th ed. Philadelphia, PA: Elsevier Saunders; chap 199.

• Busher JT. (1990). In: *Clinical Methods: The History, Physical, and Laboratory Examinations* (Walker HK, Hall WD, Hurst JW, eds) 3rd edition, Boston: Butterworths.

• Ferri C, Zignego AL, Pileri SA. (2002). Cryoglobulins. *Journal of Clinical Pathology.*55:4-13.

• Chen L, Zhao H, Yang X, Gao JY, Cheng J. (2014). HBsAg-negative hepatitis B virus infection and hepatocellular carcinoma. *Discov Med.* 18:189-93.

• Choi J, Smitz J. (2014). Luteinizing hormone and human chorionic gonadotropin: Origins of difference. *Molecular and cellular endocrinology.* 383: 203-213.

• Cole LA. (2009). New discoveries on the biology and detection of human chorionic gonadotropin. *Reprod. Biol. Endocrinol.,* 7: 8

• Cragun JM. (2011). Screening for ovarian cancer. *Cancer Control.*18:16-21.

• Daniels LB, Maisel AS. (2007). Natriuretic peptides. *J Am Coll Cardiol.* 50:2357-2368.

• Donnelly S. (2001). Why is erythropoietin made in the kidney? The kidney functions as a critmeter. *Am J Kidney Dis.* 38:415-425.

• Duffy MJ. (2001). Carcinoembryonic antigen as a marker for colorectal cancer: is it clinically useful? *Clin Chem.*47:624-630.

- El-Banawy HS, Gaber EW, Maharem DA, Matrawy KA. (2012). Angiopoietin-2, endothelial dysfunction and renal involvement in patients with systemic lupus erythematosus. *J Nephrol*. 25:541–550.

- Fall BI, Niessner R. (2009). Detection of known allergen-specific IgE antibodies by immunological methods. *Methods Mol Biol*. 509:107-122.

- Fan K, Gao L, Yan X. (2013). Human ferritin for tumor detection and therapy. *Wiley Interdiscip Rev Nanomed Nanobiotechnol*.5:287-298.

- Fang C, Zhuang Y, Wang L, Fan L, Wu Y, Zhang R, Zou A, Zhang L, Yang S, Xu W, Li J. (2013). High levels of CD20 expression predict good prognosis in chronic lymphocytic leukemia. *Cancer Science*. 104: 996-1001.

- Fejzić H, Mujagić S, Azabagić S, Burina M. (2015). Tumor marker CA 15-3 in breast cancer patients. *Acta Med Acad*. 44:39-46.

- Fischbach FT, Dunning MB III, eds. (2009). *Manual of Laboratory and Diagnostic Tests*, 8th ed. Philadelphia: Lippincott Williams and Wilkins.

- Florkowski C. (2013). HbA1c as a Diagnostic Test for Diabetes Mellitus – Reviewing the Evidence. *Clin Biochem Rev*. 34: 75-83

- González C, Aulinas A, Colom C, Tundidor D, Mendoza L, Corcoy R, Mato E, Alcántara V, Urgell Rull E, de Leiva A. (2014). Thyroglobulin as early prognostic marker to predict remission at 18-24 months in differentiated thyroid carcinoma. *Clin Endocrinol (Oxf)*. 80: 301-306.

- Goral V, Yesilbagdan H, Kaplan A, Sit D. (2007). Evaluation of CA 72-4 as a new tumor marker in patients with gastric cancer. *Hepatogastroenterology*.54:1272-1275.

- Greenfield JR, Tuthill A, Soos MA, Semple RK, Halsall DJ, Chaudhry A, O'Rahilly S. (2009). Severe insulin resistance due to anti-insulin antibodies: response to plasma exchange and immunosuppressive therapy. *Diabet Med*. 26:79-82

- Gruber HA, Farag AF. (2011). Evaluation of endocrine function. In: *Henry's Clinical Diagnosis and Management by Laboratory Methods* (McPherson RA, Pincus MR, eds.) 22nd ed. Philadelphia, Pa: Elsevier Saunders; chap 24

* Guadagni F, Roselli M, Cosimelli M, Ferroni P, Spila A, Cavaliere F, Casaldi V, Wappner G, Abbolito MR, Greiner JW. (1995). CA 72-4 serum marker--a new tool in the management of carcinoma patients. *Cancer Invest.* 13:227-238.

* Hahl P, Davis T,Washburn C, Rogers JT, Smith A. (2013). Mechanisms of neuroprotection by hemopexin: modeling the control of heme and iron homeostasis in brain neurons in inflammatory states. *J Neurochem.* 125: 89–101.

* Hande KR. (2016). Neuroendocrine tumors and the carcinoid syndrome. In: *Goldman's Cecil Medicine* (Goldman L, Schafer AI, eds.) 25th ed. Philadelphia, PA: Elsevier Saunders; chap 232.

* Hoofnagle JH, Nelson KE, Purcell RH. (2012). Hepatitis E. *N Engl J Med.* 367:1237-1244.

* Huerta MG. (2006). Adiponectin and leptin: potential tools in the differential diagnosis of pediatric diabetes? *Rev Endocr Metab Disord.* 7:187-196.

* Hyun J, Ko DH, Kang HJ, Whang DH, Cha YJ, Kim HS. (2016). Evaluation of the VIDAS Anti HCV Assay for Detection of Hepatitis C Virus Infection. *Ann Lab Med.* 36:550-554.

* Ingegnoli F, Castelli R, Gualtierotti R. (2013). Rheumatoid factors: clinical applications. Dis Markers. 35:727-734.

* Iqbal N and Iqbal N. (2014). Human Epidermal Growth Factor Receptor 2 (HER2) in Cancers: Overexpression and Therapeutic Implications. Molecular Biology International 2014: Article ID 852748, 9 pages.

* Jaffe EF, Lejtnenvi MC, Nova, FJ Mazer BD. (2011). Secondary hypogammaglobulinemia. *Immunol Allergy Clin North Am.* 21:141–163

* Jiang X, Dias JA, He X. (2014). Structural biology of glycoprotein hormones and their receptors: Insights to signaling. *Molecular and cellular endocrinology.* 382: 424-451.

* Jo JC, Ryu MH, Koo DH, Ryoo BY, Kim HJ, Kim TW, Choi KD, Lee GH, Jung HY, Yook JH, Oh ST, Kim BS, Kim JH, Kang YK. (2013). Serum CA 19-9 as a prognostic factor in patients with metastatic gastric cancer. *Asia Pac J Clin Oncol.* 9:324-330.

- Kałużna-Czaplińska J, Żurawicz E, Michalska M, Rynkowski J. (2013). A focus on homocysteine in autism. *Acta Biochim Pol.* 60:137-142.

- Kantff PW, Talcott JA. (1994). The prostate specific antigen. Its use as a tumor marker for prostate cancer. *Hematol Oncol Clin North Am.* 8:555-572.

- Kaplan EL, Rothermel CD, Johnson DR. (1998). Antistreptolysin O and anti-deoxyribonuclease B titers: normal values for children ages 2 to 12 in the United States. *Pediatrics.*10:86-88.

- Kaplan I, Bulut M, Atli A, Güneş M, Kaya MC, Çolpan L. (2015). Serum levels of carcinoembryonic antigen (CEA) in patients with bipolar disorder. *Acta Neuropsychiatr.*27:177-181.

- Katzmann JA, Kyle RA. (2006). Immunochemical characterization of immunoglobulins in serum, urine, and cerebrospinal fluid. In: *Manual of molecular and clinical laboratory immunology.* (Detrick B, ed.) 7th edn,. Washington, DC: American Society for Microbiology Press, p.88.

- Kelly D, Crotty G, O'Mullane J, Stapleton M, Sweeney B, O'Sullivan SS. (2016). The clinical utility of a low serum ceruloplasmin measurement in the diagnosis of wilson disease. *Ir Med J.* 109:341-343.

- Kim DH, Yun HY, Ryu DH, Han HS, Han JH, Yoon SM, Youn SJ. (2015). Preoperative CA 125 is significant indicator of curative resection in gastric cancer patients. *World J Gastroenterol.*21:1216-1221.

- Lakos G, Favaloro EJ, Harris EN, Meroni PL, Tincani A, Wong RC, Pierangeli SS. (2012). International consensus guidelines on anticardiolipin and anti-ß2-glycoprotein I testing: report from the 13th International Congress on Antiphospholipid Antibodies. *Arthritis Rheum.*64:1-10.

- Le TN, Nestler JE, Strauss JF, Wickham EP. (2012). Sex hormone-binding globulin and type 2 diabetes mellitus. *Trends Endocrinol Metab.* 23:32–40

- Lee HK, Kim KA, Lee JS, Kim NH, Bae WK, Song TJ. (2013). Window period of anti-hepatitis A virus immunoglobulin M antibodies in diagnosing acute hepatitis A. *Eur J Gastroenterol Hepatol.* 25:665-668.

- Li L, Dong M, Wang XG. (2016). The implication and significance of beta 2 microglubulin: A conservative multifunctional regulator. *Chinese Medical Journal* 129: 448-455

- Lihn AS, Pedersen SB, Richelsen B. (2005). Adiponectin: action, regulation and association to insulin sensitivity. *Obes Rev.* 6:13-21.

- Lin S, Sun QQ, Mao WL, Chen Y. (2016). Serum Immunoglobulin A (IgA) Level Is a Potential Biomarker Indicating Cirrhosis during Chronic Hepatitis B Infection. Gastroenterology Research and Practice 2016: Article ID 2495073, 6 pages.

- Liu C. (2012). Serum amyloid a protein in clinical cancer diagnosis. *Pathol Oncol Res.*18:117-121.

- Low ED. (2016). Nonpneumoccal streptococcal infections and rheumatic fever. In: Goldman's Cecil Medicine. (Goldman L, Schafer AI, eds.) 25th ed. Philadelphia, PA: Elsevier Saunders, chap 290.

- Maeda K, Tsutamoto T, Wada A, Hisanaga T, Kinoshita M. (1998). Plasma brain natriuretic peptide as a biochemical marker of high left ventricular end-diastolic pressure in patients with symptomatic left ventricular dysfunction. Am Heart J. 135:825-832.

- Martin DM, Vroon DH, Nasrallah SM. (1984). Value of serum immunoglobulins in the diagnosis of liver disease. *Liver.*4:214-218.

- Mazurkiewicz-Pisarek A, Płucienniczak G, Ciach T, Płucienniczak A. (2016). The factor VIII protein and its function. *Acta Biochim Pol.* 63:11-16.

- Miyoshi E, Nakano M. (2008). Fucosylated haptoglobin is a novel marker for pancreatic cancer: Detailed analyses of oligosaccharide structures. *Proteomics.* 8:3257-3262.

- Monroe DM, Hoffman M, Roberts HR. *(2010).* Molecular Biology and Biochemistry of the Coagulation Factors and Pathways of Hemostasis. *In: Williams Hematology. (*Prchal JT, Kaushansky K, Lichtman MA, Kipps TJ, Seligsohn U, eds.) *8th ed. New York:* McGraw-Hill.

- Motyckova G, Murali M. (2011). Laboratory testing for cryoglobulins. *Am J Hematol.* 86:500-502.

- Nasrullah M, Wesolowski LG, Meyer WA. (2013). Performance of a fourth-generation HIV screening assay and an alternative HIV diagnostic testing algorithm. *AIDS*. 27:731-737.

- Naz RK, Menge AC. (1994). Antisperm antibodies: origin, regulation, and sperm reactivity in human infertility. *Fertil Steril*.61:1001-1013.

- Nigro E, Scudiero O, Monaco ML, Plamieri A, Mazzarella G, Costagliola, Bianco A, Daniele A. (2014). New insight into adiponectin role in obesity and obesity related diseases. Article ID 658913, 14 pages.

- Norouzirad R, Shakurnia AH, Assarehzadegan M-A, Serajian A, Khabazkhoob M. (2014). Serum Levels of Anti-Hepatitis B Surface Antibody Among Vaccinated Population Aged 1 to 18 Years in Ahvaz City Southwest of Iran. *Hepatitis Monthly*. 14:e13625.

- Orino K, Watanabe K. (2008). Molecular, physiological and clinical aspects of the iron storage protein ferritin. *Vet J*.178:191-201.

- Pagana KD, Pagana TJ. (2010). Mosby's Manual of Diagnostic and Laboratory Tests, 4th ed. St. Louis: Mosby Elsevier.

- PaganaKD, Pagana TJ. (2011). Mosby's Diagnostic and Laboratory Test Reference 10th Edition: Mosby, Inc., Saint Louis, MO. pp 223-224.

- Pappa T, Ferrara AM, Refetoff S. (2015). Inherited defects of thyroxine-binding proteins. *Best Pract Res Clin Endocrinol Metab*.29:735-747.

- Persson PB. (2003). Renin: origin, secretion and synthesis. *J Physiol*. 552:667-671.

- Poole KE, Reeve J. (Dec 2005). Parathyroid hormone - a bone anabolic and catabolic agent. Current Opinion in Pharmacology.5 (6): 612–7.

- Ragni MV, Machin N, Malec LM, James AH, Kessler CM, Konkle BA, Kouides PA, Neff AT, Philipp CS, Brambilla DJ. (2016). Von Willebrand factor for menorrhagia: a survey and literature review. *Haemophilia*.22:397-402.

- Ratnam I, Chiu C, Kandala NB, Easterbrook PJ. (2006). Incidence and risk factors for immune reconstitution inflammatory syndrome in an ethnically diverse HIV type-1-infected cohort. *Clin Infect Dis* 42: 418-427.

- Rehfeld JF, Bardram L, Hilsted L, Poitras P, Goetze JP (2012). Pitfalls in diagnostic gastrin measurements. *Clin Chem.*58:831-836.

- Rugge JB, Bougatsos C, Chou R. (2015). Screening and treatment of thyroid dysfunction: an evidence review for the U.S. Preventive Services Task Force. *Ann Intern Med.*162:35-45.

- Salwen MJ, Siddiqi HA, Gress FG, Bowne WB. (2011). Laboratory diagnosis of gastrointestinal and pancreatic disorders. In: *Henry's Clinical Diagnosis and Management by Laboratory Methods.* (McPherson RA, Pincus MR, eds.) 22nd ed. Philadelphia, PA: Elsevier Saunders; chap 22.

- Santangelo S, Scarlata S, Poeta ML, Bialas AJ, Paone G, Incalzi RA. (2017). Alpha-1 Antitrypsin Deficiency: Current Perspective from Genetics to Diagnosis and Therapeutic Approaches. *Curr Med Chem.* 24: 65-90.

- Scarà S, Bottoni P, Scatena R. (2015). CA 19-9: Biochemical and Clinical Aspects. *Adv Exp Med Biol.* 867:247-260.

- Schaefer E, Anthanont P, Diffenderfer MR, Polisecki E, Asztalos BF. (2016). Diagnosis and treatment of high density lipoprotein deficiency. *Progress in Cardiovascular Diseases.* 59:97–106.

- Selby C. (1990). Sex hormone binding globulin: origin, function and clinical significance. *Ann Clin Biochem.* 27:532-541.

- Sen ES, Ramanan AV. (2014). How to use antistreptolysin O titre. *Arch Dis Child Educ Pract Ed.* 99:231-238.

- Sharma S, Jackson PG, Makan J. (2004). Cardiac troponins. *J Clin Pathol* 57: 1025-1026

- Shattock AG, Morris MC. (1991). Evaluation of commercial enzyme immunoassays for detection of hepatitis delta antigen and anti-hepatitis delta virus (HDV) and immunoglobulin M anti-HDV antibodies. *J Clin Microbiol.* 29: 873–1876.

- Shih AW, McFarlane A, Verhovsek M. (2014). Haptoglobin testing in hemolysis: measurement and interpretation. *Am J Hematol.* 89:443-447.

- Siotto M, Simonelli I, Pasqualetti P, Mariani S, Caprara D, Bucossi S, Ventriglia M, Molinario R, Antenucci M, Rongioletti M, Rossini PM,

Squitti R. (2016). Association Between Serum Ceruloplasmin Specific Activity and Risk of Alzheimer's Disease. *J Alzheimers Dis.* 50:1181-1189.

- Smith A, Mcculloh RJ. (2015). Hemopexin and haptoglobin: allies against heme toxicity from hemoglobin not contenders. *Frontiers in Physiology* 6: doi: 10.3389/fphys.2015.00187

- Soni S, Roneo C, Katz N, Cruz D. (2009). Early diagnosis of acute kidney injury: the promise of novel biomarkers. *Blood Purif.* 28: 165-174.

- Sporek M, Dumnicka P, Gala-Bladzinska A, Ceranowicz P, Warzecha Z, Dembinski A, Stepien E, Walocha J, Drozdz R, Kuzniewski M, Kusnierz-Cabala B. (2016). Angiopoietin-2 Is an Early Indicator of Acute Pancreatic-Renal Syndrome in Patients with Acute Pancreatitis. Mediators. *Inflamm.* 2016:5780903.

- Springer D, Jiskra J, Limanova Z, Zima T, Potlukova E. (2017). Thyroid in pregnancy: From physiology to screening. *Crit Rev Clin Lab Sci.* 54:102-116.

- Stoller JK, Lacbawan FL, Aboussouan LS. Alpha-1 Antitrypsin Deficiency (2017) In: GeneReviews® [Internet]. (Pagon RA, Adam MP, Ardinger HH, Wallace SE, Amemiya A, Bean LJH, Bird TD, Ledbetter N, Mefford HC, Smith RJH, Stephens K eds.) Seattle (WA): University of Washington, Seattle; 1993-2017.

- Szekanecz Z, Soós L, Szabó Z, Fekete A, Kapitány A, Végvári A, Sipka S, Szücs G, Szántó S, Lakos G. (2008). Anti-citrullinated protein antibodies in rheumatoid arthritis: as good as it gets? *Clin Rev Allergy Immunol.*34:26-31.

- Tawfeek MA, Alfadhli EM, Alayoubi AM, El-Beshbishy HA, Habib FA. (2017). Sex hormone binding globulin as a valuable biochemical marker in predicting gestational diabetes mellitus. *BMC Womens Health.*17:18.

- Trial J, Potempa LA, Entman ML. (2016). The role of C-reactive protein in innate and acquired inflammation: new perspectives. *Inflamm cell Signal* 3: pii: e1409

• Uehara Y, Saku K. (2014). High-density lipoprotein and atherosclerosis: Roles of lipid transporters. *World J Cardiol.* 6:1049-1059.

• Vassiliev V, Harris ZL, Zatta P. (2005). Ceruloplasmin in neurodegenerative diseases. *Brain Res Rev* 49: 633-640.

• Wang C, Wang Q, Li R, Duan JY, Wang CB. (2016). Synovial Fluid C-reactive Protein as a Diagnostic Marker for Periprosthetic Joint Infection: A Systematic Review and Meta-analysis. *Chin Med J (Engl).*129:1987-1993.

• Wasung ME, Chawla LS, Madero M. (2015). Biomarkers of renal function, which and when. *Clin Chim Acta.* 438: 350-357.

• Wong RC, Gillis D, Adelstein S, Baumgart K, Favaloro EJ, Hendle MJ, Homes P, Pollock W, Smith S, Steele RH, Sturgess A, Wilson RJ. (2004). Consensus guidelines on anti-cardiolipin antibody testing and reporting. *Pathology.* 36:63-68.

• Wu AH, Apple FS, Gibler WB, Jesse RL, Warshaw MM, Valdes R Jr. (1999). National Academy of Clinical Biochemistry Standards of Laboratory Practice: recommendations for the use of cardiac markers in coronary artery diseases. *Clin Chem.*45:1104-1121.

• Yang C, Li H, Wang Z, Zhang W, Zhou K, Meng J, Zhao Y, Pan J, Lv X, Liang H, Jiang X. (2012). Glycated albumin is a potential diagnostic tool for diabetes mellitus. *Clin Med (Lond).*12:568-571.

• Yip CH, Rhodes A. (2014). Estrogen and progesterone receptors in breast cancer. *Future Oncol.* 10: 2293-2301.

• Yosten GL, Maric-Bilkan C, Luppi P, Wahren J. (2014). Physiological effects and therapeutic potential of proinsulin C-peptide. *Am J Physiol Endocrinol Metab.* 307:E955-E968.

• Yu C, Gershwin ME, Chang C. (2014). Diagnostic criteria for systemic lupus erythematosus: a critical review. *J Autoimmun.* 48-49:10-13.

• Zijenah LS, Katzenstein DA, Nathoo KJ, Rusakaniko S, Tobaiwa O, Gwanzura C, Bikoue A, Nhembe M, Matibe P, Janossy G (2005). T lymphocytes among HIV-infected and -uninfected infants: CD4/CD8 ratio as a potential tool in diagnosis of infection in infants under the age of 2 years. *J Transl Med* 3:6.

CHAPTER 4

THERAPEUTIC PROTEINS

Proteins are known to exhibit great therapeutic potential against several diseases and syndromes. Therapeutic proteins have been shown to be effective in the treatment of many diseases such as diabetes, heart disorders and cancer (Karacali et al., 2014; Akash et al., 2015). The US Food and Drug Administration (USFDA) have approved more than 130 different proteins for clinical use and many more are in developmental stage (Leader et al., 2008). Since the list of therapeutic proteins is very vast, a concerted effort has been made to present only the important therapeutic proteins.

USFDA's Center for Drug Evaluation and Review (CDER) and the Center for Biologics Evaluation and Review (CBER) have approved 62 recombinant therapeutic proteins. These 62 proteins comprise 48 % monoclonal antibodies, 19 % coagulation factors and 11 % replacement enzymes and the remaining 22 % includes fusion proteins, hormones, growth factors and plasma proteins. More than 50 % of the approved proteins are indicated for oncology and haematology, while the remaining serves for cardiology, dermatology, immunology, ophthalmology etc (Lagasse et al., 2017).

Advantage of protein therapeutics over small-molecule drugs (Leader et al., 2008).

1. Proteins serve a highly specific and complex set of functions that cannot be mimicked by simple chemical compounds.

2. There is potential for protein therapeutics to interfere with normal biological processes and cause adverse effects.

3. Body naturally produces many of the proteins that are used as therapeutics; these agents are often well tolerated and are less likely to elicit immune responses.

4. For diseases in which a gene is mutated or deleted, protein therapeutics can provide effective replacement treatment without the need for gene therapy, which is not currently available for most genetic disorders.

5. Clinical development and FDA approval time of protein therapeutics may be faster than that of small-molecule drugs.

6. As proteins are unique in form and function, companies are able to obtain far-reaching patent protection for protein therapeutics.

Protein therapeutics are classified based on their function and therapeutic applications (Table. 4.1) (Leader et al., 2008).

Table 4.1 Therapeutic proteins

Group I	Group II	Group III	Group IV
Endocrine disorders	**Hematopoiesis**	**Cancer**	**Protection**
Insulin	Erythropoietin	Cetuximab	Hepatitis B surface antigen (Hbs Ag)
Growth hormone	Filgrastim	Rituximab	HPV vaccine
Hemostasis and Thrombosis	**Fertility**	Trastuzumab	**Autoimmune disorders**
Factor VIII	Follicle Stimulating Hormone (FSH)	**Immunoregulation**	Anti-Rhesus (Rh) immunoglobulin G
Antithrombin III	Human chorionic gonadotropin (HCG)	Etanercept	
Protein C concentrate	**Immunoregulation**	Infliximab	
von Willebrand factor	Type I alpha-interferon	Eculizumab	

Group I	Group II	Group III	Group IV
Immunodeficiencies		**Transplantation**	
Pooled immunoglobulins	Interferon alpha 2a	Antithymocyte globulin (rabbit)	
Other deficiencies	PEG Interferon alpha 2a	Basiliximab	
Human albumin	**Endocrine disorders**	**Pulmonary disorders**	
	Teriparatide	Palivizumab	
	Growth regulation	**Infectious diseases**	
	Octreotide	Enfuvirtide	
	Recombinant human bone morphogenetic protein 7	**Endocrine disorders**	
	Becaplermin	Pegvisomant	
		Other	
		Digoxin immune serum Fab	

GROUP I

These therapeutics replace a protein that is deficient or abnormal. All enzyme therapeutics coming under Group I are presented in Chapter 6.

Endocrine Disorders

Insulin   Insulin is a peptide hormone, which regulates glucose homeostasis in the body. Reduced level of this hormone leads to type 1 diabetes mellitus which at times could be lethal. Insulin extracted from pigs and cattle has long been the source of usage. Some of the side-effects that have occurred with continued long-term use of insulin could be due to the contaminating compounds present in the animal insulin. Recombinant human insulin does not appear to have such problems (Trade name: Humulin, Novolin). Patients suffering from type 1 diabetes cannot make insulin due to their damaged or destroyed beta cells in the pancreas. Therefore, they will need insulin injections to allow their body to utilize glucose and avoid complications from hyperglycemia. People with type 2 diabetes may need insulin shots to help them better utilize sugar and to prevent long-term complications (Cichocka et al., 2016).

Growth Hormone or Somatotropin   Recombinant human growth hormone (HGH) is administered for long-term treatment of pediatric patients who have growth failure due to an inadequate secretion of endogenous growth hormone or chronic renal insufficiency; and for long-term replacement therapy in adults with growth hormone deficiency (GHD) of either childhood or adult onset (Trade name: Genotropin, Humatrope). GHD adults have increased body fat and reduced muscle mass and consequently, reduced strength and exercise tolerance. In addition, they are osteopenic, have unfavourable cardiac risk factors and impaired quality of life. In these individuals, replacing GH reverses these anomalies, although it may not alter the reduced insulin-sensitivity (Capozzi et al., 2014).

Hemostasis and Thrombosis

***Factor* VIII**   Hemophilia A is a congenital bleeding disorder caused by decreased activity of plasma coagulation factor VIII, which is managed by administration of this factor (Aledort et al., 2014) [Trade name: Bioclate, Helixate]. Factor VIII (anti-hemophilic factor A) is an essential blood coagulation protein synthesized mainly by hepatocytes, but also in kidneys, endothelial cells and lymphatic tissue. The function of FVIII in the coagulation cascade is to accelerate FX activation in the presence of FIXa, phospholipids and calcium ions.

***Antithrombin* III**   Antithrombin III is an anticoagulant protein, which inhibits thrombin (Factor IIa), Factor Xa, and other serine proteases in the coagulation cascade; its anticoagulant action is dramatically enhanced by heparin. Deficiency of this protein is associated with thromboembolic disease. Antithrombin deficiency can be inherited or acquired. Hereditary antithrombin deficiencies (HATD) can be classified into two types. In type I, an overall reduction in the production of antithrombin occurs, while in type II, antithrombin deficiencies are characterized by normal levels of antithrombin with impaired functional activity. Thrombate III is a human antithrombin indicated for patients with HATD and prevention of thromboembolism. It also prevents and treats thromboembolism in patients with HATD during high-risk situations, including pregnancy and childbirth (Refaei et al., 2017).

Protein C Concentrate   Protein C concentrate [Trade name: Ceptrotin] is indicated for patients with severe congenital Protein C deficiency for the prevention and treatment of venous thrombosis and purpura fulminans. Venous thrombosis is a clot in the vein, usually in the lower extremities. Often, this clot may break and travel to the lungs leading to pulmonary embolism. Purpura fulminans is a thrombotic disorder in which there is hemorrhagic infarction of skin and disseminated intravascular coagulation (Manco-Johnson et al., 2016).

von Willebrand Factor   von Willebrand disease (vWD) is a common inherited bleeding disorder caused by the deficiency or dysfunction of von Willebrand factor (vWF). Deficiency or defect in vWF leads to low platelet adhesion to injured blood vessels and defective intrinsic coagulation owing to low plasma levels of factor VIII. Recombinant vWF [Trade name: Vonvendi] is indicated for adult patients with vWD to treat and control bleeding episode and also to prevent excessive bleeding during and after surgery (Sharma and Flood, 2017).

Immunodeficiencies

Pooled immunoglobulins   Pooled immunoglobulins are used to strengthen the body's natural defense system (immune system) to lower the risk of infection in persons with a weakened immune system. They are prepared from healthy human blood, and possess high level of antibodies, which help fight infections [Trade name: Octagam]. Pooled immunoglobulins are indicated as prophylactic treatment for patients with primary immunodeficiency syndrome (Ochs and Pinciaro, 2004).

Other Deficiencies

Human Albumin   Human albumin is prepared from human pool plasma. It is given to persons affected due to decreased production of albumin, increased loss of albumin (nephrotic syndrome), hypovolaemia, hyperbilirubinaemia etc. When administered intravenously [Trade name: Albumarc, Albumin], this will increase the circulating plasma volume by an amount approximately equal to the volume infused. The degree and duration of volume expansion depends upon the initial blood volume (Finfer et al., 2004)

GROUP II

Therapies that augment hematological and endocrine pathways and immune responses are included in this group.

Hematopoiesis

Erythropoietin (Epo)   Epo is a glycoprotein, which plays a significant role in the production of RBCs in bone marrow. Recombinant Epo [Trade name: Epogen, Procrit] is given to anaemic patients to increase the RBC production. Anaemia could be induced by many conditions viz., chronic renal failure, chemotherapy, myelodysplastic syndrome etc (Corwin et al., 2002). EPO is involved in the regulation of erythrocyte differentiation and maintenance of physiological level of circulating erythrocyte mass. It increases the reticulocyte count within 10 days of initiation, followed by increases in the RBC count, hemoglobin, and hematocrit, usually within 2 to 6 weeks. The rate of hemoglobin increase may vary depending on the dose of EPO administered.

Filgrastim (Granulocyte Colony Stimulating Factor; G-CSF) Filgrastim is a recombinant, non-pegylated G-CSF analog [Trade name: Neupogen]. Chemotherapy-induced neutropenia (CIN) is a common and serious complication of myelosuppressive chemotherapy. It is associated with significant morbidity and mortality and can increase the cost of cancer therapy. In these cases, G-CSF is necessary to restore important cells for immune function. Filgrastim is a growth factor that stimulates the production, maturation, and activation of neutrophils (a type of white blood cell). Filgrastim also stimulates the release of neutrophils from the bone marrow. In patients receiving chemotherapy, filgrastim can accelerate the recovery of neutrophils, reducing the neutropenic phase (the time in which people are susceptible to infections) (Kamioner et al., 2013).

Fertility

Follicle Stimulating Hormone (FSH) FSH is normally produced by the pituitary gland. It is one of the hormones essential for pubertal development and the function of women's ovaries and men's testes. In women, this hormone stimulates the growth of ovarian follicles in the ovary before the release of an egg from one follicle at ovulation. It also increases oestradiol production. In men, FSH acts on the Sertoli cells of the testes to stimulate sperm production (spermatogenesis). Lack or insufficiency of FSH can cause infertility or subfertility both in men and women. Exogenous supplementation of recombinant FSH [Trade name: Gona-F, Follistim] is given as a fertility medication in patients to increase the number of matured follicles and thereby oocytes (Van Wely et al., 2003).

Human Chorionic Gonadotropin (HCG) As a fertility drug, recombinant HCG [Trade name: Ovidrel] is used to enhance and trigger ovulation (Ludwig et al., 2003). LH or luteinizing hormone spikes just before ovulation and it stimulates the maturing egg (still in its follicle in the ovary) to complete the stages of growth just before ovulation. There is a surge of LH about 36 h before the follicle releases the egg. On injection of hCG, the body reacts the same way as it does to LH. So, upon injection, as long as follicles in the ovary are at the right stage of maturation, hCG triggers the eggs to go through a final growth spurt and ovulate (be released from the follicles) within 36 h.

Immunoregulation

Type 1 Alpha Interferon Type I alpha - interferon (IFN), 'interferes' with virus replication by activating numerous genes [Trade name: Infergen]. IFN aids in regulating the activity of the immune system. They are widely used in the treatment of chronic hepatitis C infection (van Zonneveld et al., 2004).

Interferon α2a (IFN-α2a) Interferon-alpha2a (IFN-α2a) is a multifunctional cytokine that has anti-viral, immunomodulatory and anti-tumor effects [Trade name: Roferon-A]. The anti-tumor effects of IFN-α2a are mediated through upregulation of surface expression of major histocompatibility complex (MHC) class I molecules, enhancement of the proliferation of type-I helper T cells (Th1) and generation of cytotoxic T lymphocytes (CTLs) in specific antitumor immune responses. IFN-α2a is indicated for hairy cell leukaemia, chronic myelogenous leukaemia, Kaposi's sarcoma, chronic hepatitis C infection etc.(Li et al., 2015).

PegInterferon α2a PegInterferon α2a is nothing but, recombinant interferon - α2a conjugated to polyethylene glycol (PEG) to increase shelf life [Trade name: Pegasis]. Peginterferon alfa-2a is used as a part of combination therapy for adults with chronic Hepatitis C who have compensated liver disease and who have not been previously treated with IFN- α. It is used alone or in combination with ribavirin (Fried et al., 2002).

Endocrine Disorders

Teriparatide Teriparatide (recombinant human parathyroid hormone) is a potent anabolic agent used in the treatment of osteoporosis [Trade name: Forteo]. Teriparatide is a portion of human parathyroid hormone

(PTH), amino acid sequence 1 through 34. Endogenous PTH is the primary regulator of calcium and phosphate metabolism in bone and kidney. Daily injections of teriparatide stimulate new bone formation leading to increased bone mineral density (Tashjian and Gagel, 2006).

Growth Regulation

Octreotide Octreotide, a potent somatostatin analogue, exerts pharmacological actions similar to the natural hormone, somatostatin [Trade name: Sandostatin]. It is an even more potent inhibitor of growth hormone, glucagon, and insulin than somatostatin. Like somatostatin, it also suppresses leuteinizing hormone (LH) response to gonadotropin releasing hormone (GnRH), decreases splanchnic blood flow, and inhibits release of serotonin, gastrin, vasoactive intestinal peptide, secretin, motilin, and pancreatic polypeptide. Octreotide has been used to treat the symptoms associated with metastatic carcinoid tumors (flushing and diarrhea), and vasoactive intestinal peptide (VIP) secreting adenomas (watery diarrhea). Octreotide substantially reduces and in many cases can normalize growth hormone and/or IGF-1 (somatomedin C) levels in patients with acromegaly (Tomassetti et al., 2000).

Recombinant Human Bone Morphogenetic Protein 7 (rhBMP7) Bone morphogenetic protein-7 (BMP-7), also known as osteogenic protein-1, is a member of the transforming growth factor - β (TGF-β) superfamily. BMP-7 [Trade name:Osteogenic protein-1] is an important promoter of osteoblast differentiation and it is necessary for the differentiation of preosteoblasts into mature osteoblasts. BMP-7, implanted with a type I collagen carrier, is a safe and effective treatment for tibial nonunions. BMP-7 is now widely used in a variety of complex orthopaedic conditions either as an adjunct or as an alternative. It is particularly useful in the nonunion of bone as an alternative to conventional autogenous bone grafting (ABG), where the use of ABG alone is not feasible and/or other alternative treatments have failed (Cecchi et al., 2016).

Becaplermin (Platelet-Derived Growth Factor; PDGF) Becaplermin has a biological activity similar to that of endogenous platelet-derived growth factor, which includes promoting the chemotactic recruitment and proliferation of cells involved in wound repair and enhancing the formation of granulation tissue [Trade name:Regranex]. This is indicated for the topical treatment of diabetic skin ulcers, especially for uninfected diabetic foot ulcers that extend into the subcutaneous tissue and have an adequate vascular supply (Fang and Galiano, 2008).

GROUP III

Protein therapeutics which are mainly used to target a molecule or cells or organism are included in this group. A wide array of monoclonal antibodies (Mabs) used to treat different clinical conditions fall under this category. Mabs are used to treat different diseases viz., cancer, rheumatoid arthritis etc and also used as immune suppressors to overcome transplantation rejection. Since Mabs are produced against specific antigens, they are highly specific.

Cancer

Cetuximab   Cetuximab Mab is used to treat colorectal, head and neck cancer [Trade name:Erbitux]. It binds to the epidermal growth factor receptor (EGFR) and impairs cell growth and proliferation. EGFR is found on the surface of many normal cells and cancer cells. EGFR, when activated during normal human cell-reproduction, enhances cell proliferation. However, if the cell is a cancer cell, it causes cancer cell proliferation. By binding to EGFR on cancer cell, cetuximab blocks epidermal growth factor (EGF) from binding to this receptor (activation). This stops the cell from continuing the pathway that promotes cell division and growth (Saltz et al., 2004).

Rituximab   Rituximab works by targeting the CD20, a transmembrane protein expressed on normal and malignant B-cells. Then, the body's natural immune defenses are recruited to attack and kill the marked B-cells. Stem cells (young cells in the bone marrow that will develop into the various types of cells) do not have the CD20 antigen. This allows healthy B-cells to regenerate after treatment. Rituximab [Trade name:Rituxan] is indicated for non-Hodgkin's lymphomas (NHL) and rheumatoid arthritis in combination with methotrexate (Keating et al., 2005).

Trastuzumab Trastuzumab, another cancer Mab, is used in the treatment of breast cancer, which expresses more amount of human epidermal growth factor receptor 2 (HER2) (Valabrega et al., 2007).The HER2 gene produces a protein receptor on the cell surface that signals normal cell growth and proliferation. Hence, binding of trastuzumab [Trade name:Herceptin] on to HER2 receptors (which is overexpressed by breast cancer cells) prevents cancer growth and slows down the progression.

Immunoregulation

Etanercept Etanercept (ETN) [Trade name:Enbrel] is the first anti-tumor necrosis factor (TNF) agent to be approved for the treatment of rheumatoid arthritis (RA). ETN not only reduces the signs and symptoms of RA, but also retards the progression of radiographic damage and improves the quality of life and function of patients. TNF is a naturally occuring cytokine produced primarily by activated macrophages and T cells and is involved in systemic inflammation. TNF mediates the action by binding to TNF receptors. Etanercept competitively inhibits the binding of TNF to cell surface TNF receptors, rendering TNF biologically inactive. Inhibition of TNF activity is indicated for autoimmune disorder such as RA, plaque psoriasis etc. (Gorman et al., 2002).

Infliximab Infliximab (Lipsky et al., 2000) is a chimeric Mab that binds and neutralizes tumor necrosis factor α (TNF α) [Trade name:Remicade]. TNF α is a primary cytokine which plays an important role in the activation of immune system and pro-inflammatory cytokines. It is a key cytokine involved in chronic inflammatory diseases. Infliximab disrupts the activation of pro-inflammatory cascade signalling. It also reduces the infiltration of inflammatory cells into sites of inflammation. Hence, neutralization of TNF α, is used in the treatment of rheumatoid arthritis, psoriasis etc.

Eculizumab Eculizumab [Trade name:Soliris] is a targeted therapy that targets and binds to the complement protein C5, and inhibits its cleavage to C5a and C5b, preventing the formation of the terminal complement complex. This interference prevents the destruction of red blood cells (hemolysis) and therefore results in stabilization of hemoglobin and a decrease in the need for blood transfusions in persons with paroxysmal nocturnal hemoglobinuria (PNH) (Hillmen et al., 2004).

Transplantation

***Antithymocyte Globulin* (ATG)** ATG is a concentrated anti-human T-lymphocyte immunoglobulin preparation derived from rabbits [Trade name:Thymoglobulin]. It is an immunosuppressive product for the prevention and treatment of acute rejection following organ transplantation. ATG reduces the host immune response against tissue transplants or organ allografts. The most common mode of action of this polyclonal antibody is via selective depletion of T-cells (Mohty 2007).

Basiliximab Basiliximab is a chimeric monoclonal antibody and functions as an IL-2 receptor antagonist [Trade name:Simulect]. Specifically it inhibits IL-2-mediated activation of lymphocytes, a critical pathway in the cellular immune response involved in allograft rejection. It is indicated as an immunosuppressive agent for renal transplantation patients. Basiliximab binds to the alpha-subunit (CD25) of the high-affinity IL-2 receptor. This inhibits IL-2 binding, which inhibits T-cell activation and prevents the body from mounting an immune response against the foreign kidney (Noyola-Villalobos et al., 2014).

Pulmonary Disorders

Palivizumab Palivizumab [Trade name:Synagis] is used for prophylaxis of respiratory diseases caused by respiratory syncytial virus (RSV). Synagis exhibits neutralizing and fusion-inhibitory activity against RSV. These activities inhibit RSV replication or spread. The Mab binds to the glycoprotein of RSV, thereby prevents its binding and uptake by host cellular receptors (Meissner and Long, 2003).

Infectious Diseases

Enfuvirtide Enfuvirtide is a 36 residue synthetic peptide that inhibits HIV-1 fusion with CD4 cells [Trade name:Fuzeon]. It is the first available fusion inhibitor for the treatment of HIV infection. Enfuvirtide binds to the first heptad-repeat (HR1) in the gp41 subunit of the viral envelope glycoprotein and prevents the conformational changes required for the fusion of viral and cellular membranes. It works by disrupting the HIV-1 molecular machinery at the final stage of fusion with the target cell, preventing uninfected cells from becoming infected (Matthews et al., 2004).

Endocrine Disorders

Pegvisomant Pegvisomant used in the treatment of acromegaly is a recombinant human growth hormone receptor antagonist [Trade name:Somavert]. Pegvisomant selectively binds to growth hormone (GH) receptors on cell surfaces, where it blocks the binding of endogenous GH, and thus interferes with GH signal transduction. Inhibition of GH action results in decreased serum concentrations of insulin-like growth factor-I (IGF-I), and IGF binding protein-3 (IGFBP-3). This reduces the symptoms of acromegaly (van der Lely et al., 2001).

Other

Digoxin Immune Serum Fab (Ovine) Digoxin Immune Serum Fab is a sheep antibody (26-10) FAB fragment from sheep immunized with the digoxin derivative digoxindicarboxymethylamine [Trade name:Digifab]. It is used as an antidote for overdose of digoxin. Digoxin, a cardiac glycoside similar to digitoxin, is used to treat congestive heart failure. The margin between toxic and therapeutic dose is meagre. DigiFab binds molecules of digoxin, making them unavailable for binding at their site of action on cells in the body. The Fab fragment-digoxin complex accumulates in the blood, from which it is excreted by the kidney. The net effect is to shift the equilibrium away from binding of digoxin to its receptors in the body, thereby reversing its effects (Antman et al., 1990).

<u>GROUP IV</u>

Protein therapeutics which exert protective action against a deleterious foreign agent is classified under this group. Vaccines form an important domain of protein therapeutics. Vaccines are biological preparations which target to develop effective immunity against disease causing or infectious microorganisms or cancer cells. Vaccines stimulate the body's immune system to recognize the foreign agent, destroy it and remember it, so that the immune system can easily recognize and destroy these foreign microorganisms that it encounters thereafter. Vaccines are raised specifically against the immunogenic protein components of microorganism, so that they develop immunity without affecting the individual. Group IV encompasses these prophylactic or therapeutic vaccines.

Protection

Hepatitis B surface Antigen **(HBsAg)**   Hepatitis B virus (HBV) vaccine is raised against its surface antigen (HBsAg), which is non-infectious in nature [Trade name:Engerix, Recombivax HB]. On administration, it induces specific humoral antibodies against HBsAg (anti-HBs antibodies). Hence, individuals treated with this vaccine, develop immunity towards HBV (Shouval 2003).

HPV Vaccine   Human papillomavirus (HPV) is the most common viral infection of the reproductive tract. Gardasil is a non-infectious recombinant, quadrivalent vaccine prepared from the highly purified virus-like particles (VLPs) of the major capsid proteins of human papillomavirus (HPV) types 6, 11, 16, and 18. Gardasil is approved for use in women aged 9-26 years for the prevention of cervical cancer and genital warts, as well as vulvar and vaginal precancerous lesions (Shi et al., 2007).

Autoimmune Disorders

Anti-Rhesus (Rh) Immunoglobulin G   Protein therapeutics which are required to overcome autoimmune problems come under this group and an important example is Anti-Rhesus (Rh) immunoglobulin G. Rhesus (Rh) negative mothers can develop Rh antibodies if they have an Rh-positive newborn, which could cause haemolytic disease of the newborn (HDN) in subsequent pregnancies. Administering anti-Rhesus D immunoglobulin G [Trade name:Rhophylac] to Rh negative women within 72 hours of giving birth to an Rh-positive baby is an effective way of preventing RhD alloimmunization and HDN. It is also given to suppress Rh immunization in Rh – negative individuals transfused with Rh – positive red blood cells (MacKenzie et al., 2004).

Rapid progress made in the field of protein therapeutics will ensure many more drugs for the future which are underway in preclinical and clinical trials. The latest advances in protein-engineering technologies have allowed drug developers and manufacturers to fine-tune and exploit desirable functional characteristics of proteins of interest while maintaining (and in some cases enhancing) product safety or efficacy or both. Also, improvements in characteristics of the existing protein therapeutics by conjugating them with drugs, nanoparticles and other targets is yet another fascinating area which has gained more focus. Nevertheless, protein therapeutics have helped to improve the quality and survival period of humans.

References

- Akash MSH, Rehman K, Tariq M, Chan S. (2015). Development of therapeutic proteins: advances and challenges. *Turkish J Biol.* 39: 1-16.

- Aledort L, Ljung R, Mann K, Pipe S. (2014). Factor VIII therapy for hemophilia A: current and future issues, *Expert Review of Hematology.* 7: 373-385.

- Antman EM, Wenger TL, Butler VP Jr, Haber E, Smith TW. (1990). Treatment of 150 cases of life threatening digitalis intoxication with digoxin-specific Fab antibody fragments. Final report of a multicentre study. *Circulation.* 81: 1744–1752.

- Capozzi A, Casa SD, Altieri B, Pontecorvi A. (2014). Low bone mineral density in a growth hormone deficient (GHD) adolescent. *Clin Cases Miner Bone Metab.* 10: 203–205.

- Cecchi S, Bennet SJ, Arora M. (2016). Bone morphogenetic protein-7: Review of signalling and efficacy in fracture healing. *Journal of Orthopaedic Translation* 4: 28-34.

- Cichocka E, Wietchy A, Nabrdalik K, Gumprecht J. (2016). Insulin therapy - new directions of research. *Endokrynol Pol.* 67: 314-324.

- Corwin HL, Gettinger A, Pearl RG, Fink MP, Levy MM, Shapiro MJ, Corwin MJ, Colton T EPO Critical Care Trials Group. (2002). Efficacy of recombinant human erythropoietin in critically ill patients: a randomized controlled trial. *JAMA*. 288: 2827–2835.

- Fang RC, Galiano RD. (2008). A review of becaplermin gel in the treatment of diabetic neuropathic foot ulcers. *Biologics*. 2: 1–12.

- Finfer S, Bellomo R, Boyce N, French J, Myburgh J, Norton R, SAFE Study Investigators. (2004). A comparison of albumin and saline for fluid resuscitation in the intensive care unit. *N Engl J Med*. 350: 2247–2256.

- Fried MW, Shiffman ML, Reddy KR, Smith C, Marinos G, Gonçales FL Jr, Häussinger D, Diago M, Carosi G, Dhumeaux D, Craxi A, Lin A, Hoffman J, Yu J. (2002). Peginterferon α-2a plus ribavirin for chronic hepatitis C virus infection. *N Engl J Med*. 347: 975–982.

- Gorman JD, Sack KE, Davis JC Jr. (2002). Treatment of ankylosing spondylitis by inhibition of tumor necrosis factor α. *N Engl J Med*. 346: 1349–1356.

- Hillmen P, Hall C, Marsh JC, Elebute M, Bombara MP, Petro BE, Cullen MJ, Richards SJ, Rollins SA, Mojcik CF, Rother RP. (2004). Effect of eculizumab on hemolysis and transfusion requirements in patients with paroxysmal nocturnal hemoglobinuria. *N Engl J Med*. 350: 552–559.

- Kamioner D, Fruehauf S, Maloisel F, Cals L, Lepretre S, Berthou C. (2013). Study design: two long-term observational studies of the biosimilar filgrastim Nivestim (Hospira filgrastim) in the treatment and prevention of chemotherapy-induced neutropenia. BMC Cancer 13: 547.

- Karacali S, Izzetoglu S, Deveci R. (2014). Glycosylation changes leading to the increase in size on the common core N-glycans, required enzymes and related cancer associated proteins. *Turkish J Biol*. 38: 754-771.

- Keating M J, O'Brien S, Albitar M, Lerner S, Plunkett W, Giles F, Andreeff M, Cortes J, Faderl S, Thomas D, Koller C, Wierda W, Detry MA, Lynn A, Kantarjian H. (2005). Early results of a chemoimmunotherapy regimen of fludarabine, cyclophosphamide, and rituximab as initial therapy for chronic lymphocytic leukemia. *J Clin. Oncol.* 23: 4079–4088.

- Lagassé HA, Alexaki A, Simhadri VL, Katagiri NH, Jankowski W, Sauna ZE, Kimchi-Sarfaty C. (2017). Recent advances in (therapeutic protein) drug development. 6: 113.

- Leader B, Baca QJ, Golan DE. (2008). Protein therapeutics: a summary and pharmacological classification. *Nat Rev Drug Discov.* 7: 21-39.

- Li W, Hao Q, He L, Meng J, Li M, Xue X, Zhang C, Li H, Zhang W, Zhang Y. (2015). Recombinant IFN-α2a-NGR exhibits higher inhibitory function on tumor neovessels formation compared with IFN-α2a in vivo and in vitro. *Cytotechnology.* 67: 1039–1050.

- Lipsky PE, van der Heijde DM, St Clair EW, Furst DE, Breedveld FC, Kalden JR, Smolen JS, Weisman M, Emery P, Feldmann M, Harriman GR, Maini RN; Anti-Tumor Necrosis Factor Trial in Rheumatoid Arthritis with Concomitant Therapy Study Group. (2000). Infliximab and methotrexate in the treatment of rheumatoid arthritis. Anti-Tumor Necrosis Factor Trial in Rheumatoid Arthritis with Concomitant Therapy Study Group. *N Engl J Med.* 343: 1594–1602.

- Ludwig M, Doody KJ, Doody KM. (2003). Use of recombinant human chorionic gonadotropin in ovulation induction. *Fertil. Steril.* 79: 1051–1059.

- MacKenzie IZ, Bichler J, Mason GC, Lunan CB, Stewart P, Al-Azzawi F, De Bono M, Watson N, Andresen I. (2004). Efficacy and safety of a new, chromatographically purified rhesus (D) immunoglobulin. *Eur J Obstet Gynecol Reprod Biol.* 117: 154–161.

- Manco-Johnson MJ, Bomgaars L, Palascak J, Shapiro A, Geil J, Fritsch S, Pavlova BG, Gelmont D. (2016). Efficacy and safety of protein C concentrate to treat purpura fulminans and thromboembolic events in severe congenital protein C deficiency. *Thromb Haemost.* 116: 58-68.

- Matthews T, Salgo M, Greenberg M, Chung J, DeMasi R, Bolognesi D. (2004). Enfuvirtide: the first therapy to inhibit the entry of HIV-1 into host CD4 lymphocytes. *Nature Rev Drug Discov.* 3: 215–225.

- Meissner HC, Long SS. (2003). Revised indications for the use of palivizumab and respiratory syncytial virus immune globulin intravenous for the prevention of respiratory syncytial virus infections. *Pediatrics*. 112: 1447–1452.

- Mohty M. (2007). Mechanisms of action of antithymocyte globulin: T-cell depletion and beyond. *Leukemia*. 21: 1387-1394.

- Noyola-Villalobos H, Espinoza-Mercado F, Jimenez-Chavarria E, Ramos-Diaz E, Loera-Torres M, Rivera-Navarrete E. (2014). Basiliximab Induction Therapy in Renal Transplantation: A 12-Year Retrospective Study at a Mexican Single-Transplant Center. *Transplantation*. 98: 256-257.

- Ochs HD, Pinciaro PJ. (2004). Octagam 5%, an intravenous IgG product, is efficacious and well tolerated in subjects with primary immunodeficiency diseases. *J Clin Immunol*. 24: 309–314.

- Refaei M, Xing L, Lim W, Crowther M, Boonyawat K. (2017). Management of Venous Thromboembolism in Patients with Hereditary Antithrombin Deficiency and Pregnancy: Case Report and Review of the Literature. Case Rep Hematol 2017: 9261351.

- Saltz, LB Meropol NJ, Loehrer PJ Sr, Needle MN, Kopit J, Mayer RJ. (2004). Phase II trial of cetuximab in patients with refractory colorectal cancer that expresses the epidermal growth factor receptor. *J. Clin. Oncol*. 22: 1201–1208.

- Sharma R, Flood VH. (2017). Advances in the diagnosis and treatment of Von Willebrand disease. *Blood*. 130: 2386-2391.

- Shi L, Sings HL, Bryan JT, Wang B, Wang Y, Mach H, Kosinski M, Washabaugh MW, Sitrin R, Barr E. (2007). Gardasil: prophylactic human papillomavirus vaccine development — from bench top to bedside. *Clin Pharm Ther*. 81: 259–264

- Shouval D. (2003). Hepatitis B Vaccines. *J Hepatol*. 39: S70–S76.

- Tashjian AH Jr., Gagel RF. (2006). Teriparatide [human PTH(1–34)]: 2.5 years of experience on the use and safety of the drug for the treatment of osteoporosis. *J Bone Miner Res*. 21: 354–365.

- Tomassetti P, Migliori M, Caletti GC, Fusaroli P. (2000). Treatment of type II gastric carcinoid tumors with somatostatin analogues. *N Engl J Med.* 343: 551–554.

- Valabrega GM, Montemurro F, Aglietta M. (2007). Trastuzumab: mechanism of action, resistance and future perspectives in HER2-overexpressing breast cancer. *Ann Oncol.* 18: 977–984.

- Van der Lely AJ, Hutson RK, Trainer PJ, Besser GM, Barkan AL, Katznelson L, Klibanski A, Herman-Bonert V, Melmed S, Vance ML, Freda PU, Stewart PM, Friend KE, Clemmons DR, Johannsson G, Stavrou S, Cook DM, Phillips LS, Strasburger CJ, Hackett S, Zib KA, Davis RJ, Scarlett JA, Thorner MO. (2001). Long-term treatment of acromegaly with pegvisomant, a growth hormone receptor antagonist. *Lancet.* 358: 1754–1759.

- Van Wely M, Westergaard LG, Bossuyt PM, Van der Veen, F. (2003). Human menopausal gonadotropin versus recombinant follicle stimulation hormone for ovarian stimulation in assisted reproductive cycles. *Cochrane Database Syst. Rev.* 1: CD003973.

- Van Zonneveld M, Honkoop P, Hansen BE, Niesters HG, Darwish Murad S, de Man RA, Schalm SW, Janssen HL. (2004). Long-term follow-up of α-interferon treatment of patients with chronic hepatitis B. *Hepatology.* 39: 804–810.

CHAPTER 5

ENZYMES IN CLINICAL MEDICINE

The global enzyme market size of USD 8.2 billion in 2015 is expected to reach USD 17.5 billion by 2024, according to Grand View Research, Inc. Among the enzyme products, enzymes used in medical applications are expected to achieve significant growth. "Diagnostic enzymes" refers to enzymes used for diagnosis or prognosis. Enzymes, like other proteins, are synthesized by body tissues to meet their metabolic needs. They are not always tissue-specific. More than one tissue/organ may synthesize one or more enzymes. Enzymes are preferred in diagnostics because of their substrate specificity and their quantitated activity in the presence of other proteins. Details of enzymes used in diagnosis are shown in Table 5.1.

Table 5.1 Enzymes used in diagnosis

Cardiac markers	Liver markers	Pancreatic Markers	Bone Markers	Renal markers	Specific markers	Cancer Markers
• Creatine kinase - MB	• Alkaline phosphatase	• Amylase	• Tartrate-resistant acid phosphatase	• N-acetyl beta-D-glucosamini-dase	• G6PDH	• Cathepsin D
• Glycogen phosphorylase - BB	• Alanine transaminase	• Lipase		• Leukocyte esterase	• Cholines terases	• Cysteine cathepsins
• Lactate dehydrogenase	• Aspartate transaminase			• Lysozyme		• Gelatinase B (MMP-9)
	• Gamma glutamyl transferase					• Acid phosphatase

CARDIAC MARKERS

Creatine Kinase – MB

Creatine kinase (CK) is an intracellular enzyme present in high concentrations in skeletal muscle, myocardium and brain. CK is a dimeric molecule composed of two subunits designated M and B. Combinations of these subunits form the isoenzymes CK–MM, CK–MB, and CK–BB. When there is muscle damage, CK is released from muscle cells and it is detectable in the blood. While total CK is recognized as nonspecific, CK-MB is the most specific, accurate and cost-effective means of detecting myocardial infarction (MI). CK-MM and CK-BB are found in skeletal muscle and brain, respectively. CK-MB is normally undetectable or very low in the blood. It invariably shows a rapid increase in serum of patients in the early hours after admission with chest pain. Use of CK-MB as a marker helps to diagnose MI within 6 h of the onset of symptoms and >60% of infarctions are detectable within the first hour of admission in the cardiac intensive care unit (Rosalki et al., 2004). Apart from MI, kidney failure, low thyroid hormone levels and alcohol abuse can also increase CK-MB. During recent years, CK-MB activity assays have been replaced by CK-MB mass assays which measure the protein concentration of CK-MB, rather than its catalytic activity. Enzyme immunoassays have become the choice for measuring CK-MB in the laboratory because analytical interferences which lead to false positive test results are less frequent (Mair et al., 1997).

Glycogen Phosphorylase BB (GPBB)

Glycogen phosphorylase is a glycolytic enzyme which plays an essential role in the regulation of carbohydrate metabolism by mobilization of glycogen (Rabitzsch et al., 1995). Among the three isoenzymes viz., glycogen phosphorylase LL (GPLL) [liver], glycogen phosphorylase MM (GPMM) [muscle] and glycogen phosphorylase BB (GPBB) [Brain], the isoenzyme BB is the predominant isoenzyme in the myocardium (Mair et al., 1997). A rapid rise in blood levels of GPBB can be seen in MI and unstable angina. GPBB is found to increase between 1 h and 4 h after the onset of chest pain in MI patients. Currently, GPBB is not recommended as a diagnostic marker of MI (Dobric et al., 2015).

Lactate Dehydrogenase (LDH)

LDH is a tetrameric enzyme composed of two subunits, M and H, resulting in five isozymes, the two homotetramers M4 and H4, and three hybrid forms. LDH-M predominates in muscle, while LDH-H predominates in heart and in other tissues. LDH is found in a variety of body tissues and it is released from the cells into the serum when cells are damaged or destroyed. Isoenzyme fractions are used to identify the type of tissue damage. LDH_1 and LDH_2 are found in the heart, brain, and erythrocytes. LDH_3 is found in the brain and kidneys. LDH_4 is found in the liver, skeletal muscle, and kidneys. LDH_5 is found in the liver, skeletal muscle, and ileum. LDH_2 normally accounts for the highest percentage of total serum LDH. A significant increase in the serum level of LDH-1 indicates that heart muscle has been damaged as in MI (Mair et al., 1997). LDH increases within about 12 h after MI, peaks between 24 and 48 h and normalizes by about day 10. Mair et al., (1997) have reported that liver disease is indicated when there is an elevation in the level of LDH-5. LDH levels are also elevated in conditions viz., kidney disease, pancreatitis, lymphoma, fractures, meningitis etc. LDH alone or in combination with other tumor markers or other factors may be used for the identification of recurrence following 'curative' primary therapy in cancer patients. LDH is perhaps the most common enzyme used for prognosis. It is a validated prognostic marker for germ cell malignancy of testis, lymphoma, leukemia and colon cancer, and it is also a key for monitoring patients during treatment and on surveillance. Patients can be stratified into treatment protocols based on LDH activity (Schwartz, 1992). Radenkovic et al., (2013) further adds that activity of LDH in tumor tissue along with mammographic characteristics could help in defining aggressive breast cancers. Low levels of LDH are sometimes seen when someone ingests large amounts of ascorbic acid (vitamin C) (Rama Mani et al., 2006).

LIVER MARKERS

Alkaline Phosphatase (ALP)

ALP is found in several tissues throughout the body, wherein the highest concentrations are present in the cells that comprise bone and liver. ALP, functioning under alkaline pH, is classified under hydrolases and it removes phosphate groups from nucleotides and proteins. ALP levels increase due to liver disorders (viz., hepatitis, liver cancer, cirrhosis etc) or bone disorders. The increase in the level of serum ALP also indicates an increased hepatocytic activity in hepatobiliary disease. Higher ALP levels in serum are observed when bile ducts are blocked as in the case of obstructive jaundice (Corathers, 2006). The increased levels should be correlated with clinical signs and symptoms to assess liver or bone damage. If it is not apparent, a test for ALP isoenzymes can be performed to determine the source. In case of liver diseases, ALP elevation has to be evaluated along with other tests viz., AST or ALT.

In children, serum ALP levels are considerably higher than in adults and it correlates with the rate of bone growth. ALP values are slightly higher in men than in women, but after age 60, the enzyme value is equal or higher in women. ALP concentrations are increased during puberty, pregnancy and after menopause (Corathers, 2006). The increase in the level of serum ALP indicates an increased osteoblastic activity or when there is active bone formation as in the case of PAGET's disease or rheumatoid arthritis. Other pathological conditions that may result in high levels of ALP also include rickets, osteomalacia, hyperthyroidism and hyperparathyroidism. Low activities of ALP are far less common and are more likely related to a genetic condition or nutritional deficiency. Hypophosphatasia is a rare disorder characterized by low levels of serum ALP activity resulting in abnormal phosphorylated metabolites and varying skeletal abnormality (Iqbal, 1998). Delmas (1993) has observed that determination of bone specific ALP in serum is a useful parameter for monitoring changes in bone formation and patients with bone metastases show an increased activity of this isoenzyme.

Alanine Transaminase (ALT)

ALT is mostly found in the cells of liver and kidney. Its major function is to convert alanine into pyruvate, an important intermediate in cellular energy production. Elevated ALT (also known as serum glutamate pyruvate transaminase [SGPT]) levels are also associated with an increased risk of hepatocellular carcinoma (Ishiguro et al., 2009). Increased serum level of ALT indicates a severe liver disease, usually viral hepatitis and toxic liver necrosis. Kim et al., (2004) have reported that ALT is a common serum marker of liver disease. Even a minor elevation of ALT is a good indicator of severity in liver disease.

Aspartate Aminotransferase (AST)

AST, also known as serum glutamate oxaloacetate transaminase (SGOT), is a pyridoxal phosphate (PLP) dependent enzyme. It catalyses the interconversion of aspartate and α-ketoglutarate to oxaloacetate and glutamate. AST is found in cells throughout the body but mostly in the heart and liver and to a lesser extent in the kidneys and muscles. In healthy individuals, levels of AST in the blood are low. Significant increase in the serum level (10-100 times normal) of AST indicates severe damage to liver (viral hepatitis or toxic liver necrosis) or heart cells (MI). Lesmana et al., (2011) have reported that AST to platelet ratio index (APRI) could be a much cheaper alternative and a useful marker to screen liver fibrosis in the primary care setting when transient elastography is not available. AST test is often performed along with ALT test. Though the elevation of both the enzymes indicate liver damage, a AST/ALT ratio helps to differentiate between different causes of liver injury and in recognizing whether the increased levels may be coming from alternate source, such as heart or muscle injury. Usually, in alcohol induced liver damage, cirrhosis and liver tumors, the level of AST is higher than ALT.

Gamma Glutamyltransferase (GGT)

GGT catalyzes the transfer of amino acids from one peptide to another amino acid or peptide. This enzyme is sometimes referred to as a "transpeptidase".

Specifically, it catalyzes the transfer of a gamma glutamyl group to another acceptor. GGT is found in many organs throughout the body, but, a high concentration is noticed in the liver. Normally, GGT is present in low levels, but its elevated level in blood indicates damage to the liver or bile ducts. It is the most sensitive liver enzyme used for detecting bile duct problems. Hepatobiliary disease is the predominant cause of increased serum GGT activity. Increases are associated with all forms of primary and secondary hepatobiliary disorders. Elevations are moderate (2 to 5 fold) with diffuse hepatic cell injury due to toxic or infectious hepatitis. Cholestasis due to intrahepatic or extrahepatic biliary obstruction causes higher serum levels (5 to 30 times normal level). Increases occur earlier and persist longer than ALP in cholestatic disorders (Vroon and Israili, 1990).

PANCREATIC MARKERS

Amylase

Amylase is an enzyme produced in the pancreas and the salivary glands and it converts starch, glycogen and related polysaccharides into simple and easily digested sugar. When cells in the pancreas are injured, due to pancreatitis or when the pancreatic duct is blocked by a gallstone or by a pancreatic tumor in rare cases, increased amounts of amylase are released into the blood. It is reported that elevation of total serum amylase is a sensitive marker of acute pancreatitis, within 24 h of onset of symptoms. However, after the first hospital day, it is the least sensitive of the enzymatic tests. Shen et al., (2012) have reported that high initial salivary α amylase activity is an independent predictor of acute MI in patients brought to the emergency ward with chest pain.

Lipase

Lipase is an important digestive enzyme secreted by the pancreas, which aids in the breakdown of dietary triglycerides into fatty acids. Increased levels indicate pancreatic damage or block in the pancreatic duct. Level of lipase in serum can be used as a diagnostic tool for detecting conditions such as acute pancreatitis and pancreatic injury (Suzuki et al., 2014). A low level of lipase may indicate a damage to the lipase producing cells in the pancreas, cystic fibrosis etc.

BONE MARKERS

Tartrate-Resistant Acid Phosphatase (TRAP)

TRAP is a glycosylated monomeric metalloenzyme expressed in mammals, with a molecular weight of approximately 35 kDa. It is differentiated from other mammalian acid phosphatases by its resistance to inhibition by tartrate, molecular weight and characteristic purple colour. TRAP is an osteoclast enzyme, used as a marker of bone resorption. TRAP is expressed by osteoclasts, macrophages, dendritic cells and a number of other cell types. The fraction of TRAP that is most specific of the osteoclasts is the tartrate-resistant acid phosphatase isoform 5b (TRACP5b). TRAP level is increased in rheumatoid arthritis, osteoporosis and metabolic bone disorders. TRACP 5b is a biomarker for diagnosis as well as prognosis in various cancers with high incidence of bone metastasis including breast, prostate, lung, and multiple myeloma (Chao et al., 2010).

RENAL MARKERS

N-acetyl-β-D-glucosaminidase (NAG)

Gatsing et al., (2006) have investigated three urinary lysosomal glycosidases viz., N-acetyl-β-D-glucosaminidase [NAG] (β-hexosaminidase), β-glucuronidase and β-galactosidase and they are found to be of particular diagnostic value in the early detection of diabetic nephropathy, with NAG being the most useful indicator. Karakani et al., (2007) also state that measurement of urinary NAG in diabetic patients serves as a biomarker for screening diabetic renal dysfunction. Urinary NAG excretion is a marker of tubular cell dysfunction and a predictor of outcome in primary glomerulonephritis. Urinary NAG excretion can be considered as a reliable marker of the tubulo-toxicity of proteinuria in the early stage of idiopathic membranous nephropathy (IMN), primary focal segmental glomerulosclerosis (FSGS) and minimal change disease (MCD). NAG determination may be non-invasive, useful test for the early identification of patients who will subsequently develop chronic renal failure or clinical remission and responsiveness to therapy (Bazzi et al., 2002).

Leukocyte Esterase

Leukocyte esterase is produced by the leukocytes (WBC). A negative test is normal and its presence in urine indicates urinary tract infection. Parvizi et al., (2011) have stated that detection of leukocyte esterase in synovial fluid is an extremely valuable addition to the physician's armamentarium for the diagnosis of periprosthetic joint infection.

Lysozyme

Lysozyme is widely distributed in a variety of tissues and body fluids including liver, articular cartilage, plasma, saliva, tears and milk. It possesses anti-bacterial activity especially against gram-positive bacteria and hence plays a significant role in host defense. Lysozyme hydrolyses glycosidic bonds in the cell wall of peptidoglycans. In the case of normal subjects, at least 500 mg of lysozyme is produced per day, but the lifetime of the protein in plasma is very short; ~ 75% is eluted within 1 h, mainly through clearance by the kidneys. Normal values are in the range of 5–11 µg/mL, while elevated levels indicate granulocytic or lymphocytic leukemia. Torsteinsdottir et al., (1999) have shown that serum levels of lysozyme are elevated in rheumatoid arthritis patients, which serve as an indicator of monocyte/macrophage activity. Serum lysozyme levels may also be elevated in tuberculosis, bacterial infections, urinary tract infections, glomerulonephritis etc.

SPECIFIC MARKERS

Glucose-6-Phosphate Dehydrogenase (G6PD)

Glucose-6-phosphate dehydrogenase (G6PD), an enzyme involved in energy production, is found in all cells, including red blood cells (RBCs) and it protects them from deleterious effects of reactive oxygen species (ROS). G6PD, in the RBC hexose monophosphate pathway, acts to produce reduced glutathione. Subsequently, glutathione prevents oxidative damage to hemoglobin and other intracellular structures. Thus, in G6PD deficiency, hemoglobin becomes oxidized and gets precipitated inside RBCs leading to hemolytic anemia; precipitated hemoglobin inside RBCs is identified by light microscopy as 'Heinz Bodies.' G6PD deficiency is a genetic disorder, which can cause persistent jaundice in newborns. If left untreated, this can lead to significant brain damage and mental retardation. Wang et al., (2012) report that overexpression of G6PD is closely related to progression of gastric cancer and might be regarded as an independent predictor of prognosis for gastric cancer. Devi et al., (1997) have demonstrated that leukocyte G6PD could serve as a diagnostic and prognostic tool in acute non-lymphocytic leukemia (ANLL) and chronic myeloid leukemia (CML). The above study reports that G6PD level has been found to be significantly decreased in majority of the patients with ANLL while it is increased in all CML patients.

Cholinesterases

Cholinesterases are a family of enzymes present in the central nervous system, particularly in nervous tissue, muscle and red blood cells and they catalyze the hydrolysis of the neurotransmitter acetylcholine into choline and acetic acid. There are two types viz., acetylcholinesterase (AChE) and pseudocholinesterase or butyrylcholinesterase (BChE). AChE is primarily found on red blood cell membranes, in neuromuscular junctions, and in neural synapses, while BChE is produced in the liver and found primarily in plasma. The difference between the two types of cholinesterases is their relative preferences for substrates. AChE hydrolyzes acetylcholine faster while BChE hydrolyzes butyrylcholine faster. Decreased levels of cholinesterase indicate malnutrition, liver disease, poisoning from organophosphates, renal disease etc. Mabrouk et al., (2011) have shown that patients with schizophrenia have higher plasma BChE activity than controls.

CANCER MARKERS

Cathepsin D

Cathepsin D is an aspartic endopeptidase which plays an important role in immunity, tissue remodelling, adipogenesis etc. Cathepsin D is synthesized in rough endoplasmic reticulum as preprocathepsin D. After removal of signal peptide, the 52 kDa procathepsin D is targeted to intracellular vesicular structures (lysosomes, endosomes, phagosomes) (Kornfeld, 1990). Cathepsin D is present in all tissues, specifically in kidneys, spleen, liver and macrophages. The activity of cathepsin D causes the proteolysis of cellular and extracellular proteins. Considerable intensity of cellular degradation and increased activity of cathepsin D are observed in some inflammations. Sohar et al., (2002) report that elevated level of lysosomal cathepsin D is associated with progression of rheumatoid arthritis. Increased activity has also been observed in the course of various neoplastic diseases. Schwartz (1992) reports that cathepsin D in breast tissue may be useful in predicting women with breast cancer who are at risk for early recurrence. The average activity of cathepsin D in normal individuals is 11.8 nM (10^{-2} nM tyrosine/mg protein/min) (Kuzniewski, 2014).

Cysteine Cathepsins (CCs)

CCs are important enzymes in cancer and they are involved in apoptosis, angiogenesis, cell proliferation and invasion. In normal cells, CCs are localized in lysosomes and other intracellular compartments and they are involved in protein degradation and processing. However, in certain tumors, they are translocated from their intracellular compartments to the cell surface and are even secreted. The best known CCs, cathepsins B, L and H are distributed ubiquitously and they catalyze protein hydrolysis within the lysosomes. Enhanced expression of CCs has been demonstrated in many human tumors, including breast, ovary, uterine cervix, lung, brain, gastrointestinal, head, neck and melanoma. Upregulation of cathepsin B has been observed in premalignant lesions in colon, thyroid, brain, liver, breast and prostate (Koblinski et al., 2000). CCs have also been implicated in inflammatory diseases such as inflammatory myopathies, rheumatoid arthritis, and periodontitis. Hence, CCs play a potential role as diagnostic and prognostic marker in cancer as well as in certain inflammatory disorders (Berdowska, 2004). Hersze´nyic et al., (2000) have observed that cysteine and serine proteases may also have a role as tumour markers in the early diagnosis of gastrointestinal tract tumours.

Gelatinase B

Gelatinase B (also called as Matrix metalloproteinase-9 [MMP-9]) is a secreted enzyme that regulates cell-matrix composition. It belongs to the gelatinase subfamily of the MMPs and therefore, its main substrate is gelatin (a denatured collagen). MMP-9 is produced by selected cell types including keratinocytes, monocytes, tissue macrophages, polymorphonuclear leukocytes and a variety of malignant cells. MMP-9 appears to be involved in a variety of pathological processes that occur in autoimmune diseases (Ram et al., 2006). Elevated levels are also observed in pancreatic cancer, breast cancer etc. Its increase is also observed in saliva in case of periodontal diseases and gingivitis. Blanco-Prieto et al., (2017) have observed that MMP-9 is a potential biomarker for non-small cell lung cancer (NSCLC) detection and its combined measurement with other biomarkers could improve NSCLC detection.

Acid Phosphatase (ACP)

ACP is a lysosomal enzyme which is present primarily in bone, prostate, platelets, erythrocytes and spleen. Prostate acid phosphatase (PACP) has been used extensively as a serum marker for cancer of the prostate. ACP level in male prostate gland is 100 times more than in any other body tissue. ACP assay is carried out to check whether prostate cancer has metastasized and also to monitor the prognosis. It has also been reported that increased ACP activity is observed in the serum and tissues of patients with Gaucher's disease, an inborn error of cerebroside metabolism.

Table 5.2   Normal and abnormal levels of serum enzymes in diagnosis

Enzyme	Normal level		Abnormal level	Associated disease or disorder
CARDIAC MARKERS				
CK-MB		< 20 IU/L	High	Myocardial infarction
LDH	Child	60 – 120 U/L	High	Myocardial infarction
	Adult	140 – 280 U/L		
LIVER MARKERS				
Alkaline phosphatase		20 – 140 IU/L	High	Liver cirrhosis, liver cancer, hepatitis, osteoblastic bone tumors, etc
			Low	Malnutrition, hypophosphatasia etc
Alanine transaminase (SGPT)		5 – 40 U/L	High	Liver diseases
Aspartate aminotransferase (SGOT)		5 – 35 U/L		Liver disease, heart disease, muscle injury
GGT		< 50 IU/L	High	Liver disease, bile duct obstruction etc
PANCREATIC MARKERS				
Amylase	Serum	25 – 100 IU/L	High	Pancreatitis, pancreatic tumor
			Low	Kidney disease, damage to amylase producing cells in pancreas

Enzyme	Normal level		Abnormal level	Associated disease or disorder
Lipase		25 – 100 U/L	High	Pancreatitis
			Low	Cystic fibrosis
BONE MARKERS				
TRAP		2.5-5.0 ng/mL	High	Rheumatoid arthritis, bone disorders, Gaucher's disease
RENAL MARKERS				
N-acetyl β-D glucosaminidase	Urine	1.6-5.8 U/g creatinine	High	Kidney disorder
Leukocyte esterase		Absent	Present	Urinary tract infection
Lysozyme		5 – 11 µg/mL	High	Rheumatic arthiritis, Granulocytic or lymphocytic leukemia, bacterial infections, renal dysfunction etc
SPECIFIC MARKERS				
Glucose – 6 - phosphate dehydrogenase (G6PD)		5.5 – 20.5 Units/g of hemoglobin	< 10 % of normal	Chronic hemolytic anemia
			10 – 60 % of normal	Intermittent anemia
Butyrylcholinesterase (Pseudocholinesterase)	Male	3100 – 6500 U/mL	High	Schizophrenia
	Female	1800 – 6800 U/mL	Low	Organophosphate toxicity, liver disease etc

Enzyme	Normal level	Abnormal level	Associated disease or disorder
CANCER MARKERS			
Cathepsin D	11.8 nM	High	Rheumatoid arthritis, inflammation, breast cancer
Cysteine cathepsins	Absent	Present	Cancer
Gelatinase B	11-59 ng/mL	High	Rheumatoid arthritis, cancer, myocardial infarction
Acid phosphatase	Upto 3 IU/L	High	Prostate cancer

References

- Bazzi C, Petrini C, Rizza V, Arrigo G, Napodano P,Paparella M, Damico G. (2002). Urinary N-Acetyl Beta-D-glucosaminidase excretion is a marker of tubular cell dysfunction and a predictor of outcome in primary glomerulonephritis. *Nephrol Dial Transplant.* 17: 1890-1896.

- Berdowska I. (2004). Cysteine proteases as disease markers. *Clin Chim Acta.* 342: 41-69.

- Blanco-Prieto S, Barcia-Castro L, Cadena M, Rodriguez-Berrocal F, Vazquez-Iglesias L, Botana-Rial MI, Fernández-Villar A, De Chiara L. (2017). Relevance of matrix metallo proteases in non-small cell lung cancer diagnosis. *BMC cancer.* 17: 823.

- Chao T, Yi-Ying Wu, Janckila AJ. (2010). Tartrate-resistant acid phosphatase isoform 5b (TRACP 5b) as a serum maker for cancer with bone metastasis. *Clin Chim Acta.* 411: 1553-1564.

- Corathers SD. (2006). The alkaline phosphatase level: nuances of a familiar test, *Pediatr Rev.* 27: 382-384.

- Delmas PD. (1993). Biochemical markers of bone turnover I: theoretical considerations and clinical use in osteoporosis. *Am J Med.* 95: 11S-16S.

- Devi GS, Prasad MH, Reddy PP, Rao DN. (1997). Leukocyte glucose 6 phosphate dehydrogenase as prognostic indicator in ANLL and CML. *Indian J Exp Biol.* 35: 155-158.

- Dobric M, Ostojic M, Giga V, Djordjevic-Dikic A, Stepanovic J, Radovonovic N, Beleslin B. (2015). Glycogen phosphorylase BB in myocardial infarction. *Clin Chim Acta.* 438: 107-111.

- Gatsing D, Garba IB, Adoga GI. (2006). The use of lysosomal enzymuria in the early detection and monitoring of the progression of diabetic nephropathy. *Indian J Clin Biochem,* 21: 42-48.

- Hersze´nyic L, Plebania M, Carrarob P, De Paoli M, Roveroni G, Cardin R, Foschia F, Tulassay Z, Naccarato R, Farinatia F. (2000). Proteases in gastrointestinal neoplastic diseases. *Clin Chim Acta.* 291: 171-187.

- Iqbal SJ. (1998). Persistently raised serum acid phosphatase activity in a patient with hypophosphatasia: electrophoretic and molecular weight characterisation as type 5, *Clin Chim Acta.* 271: 213-220.

- Ishiguro S, Inoue M, Tanaka Y, Mizokami M, Iwasaki M, Tsugane S. (2009). Serum aminotransferase level and the risk of hepatocellular carcinoma: a population-based cohort study in Japan. *Eur J Cancer Prev.* 18: 26-32.

- Karakani AM, Haghighi S A, Khansari MG, Hosseini R. (2007). Determination of urinary enzymes as a marker of early renal damage in diabetic patients. *J Clin Lab Anal.* 21: 413-417.

- Kim HC, Nam CM, Jee SH, Han KH, Oh DK, Suh I. (2004). Normal serum aminotransferase concentration and risk of mortality from liver diseases: prospective cohort study. *Br Med J.* 328: 983.

- Koblinski JE, Ahram M, Sloane BF. (2000). Unraveling the role of proteases in cancer. *Clin Chim Acta.* 291: 113-135.

- Kornfeld S. (1990). Lysosomal enzyme targeting. *Biochem Soc Trans.* 18: 367-374.

- Kuźniewski R. (2014). Activity of cathepsin D and a1-antitrypsin in blood of men with malignant melanoma. *Postep Derm Alergol.* 31:170–173.

- Lesmana CR, Salim S, Hasan I, Sulaiman AS, Gani RA, Pakasi LS, Lesmana LA, Krisnuhoni E, Budihusodo U. (2011). Diagnostic accuracy of transient elastography (FibroScan) versus the aspartate transaminase to platelet ratio index in assessing liver fibrosis in chronic hepatitis B: the role in primary care setting. *J Clin Pathol.* 64: 916-920.

- Mabrouk H, Mechria H, Mechri A, Rahali H, Douki W, Gaha L, Fadhel Najjar M. (2011), Butyrylcholinesterase activity in schizophrenic patients. *Ann Biol Clin* (Paris), 69: 647-652.

- Mair J. (1997). Cardiac troponin I and troponin T: are enzymes still relevant as cardiac markers? *Clin Chim Acta.* 257: 99-115.

- Parvizi J, Jacovides C, Antoci V, Ghanem E. (2011). Diagnosis of periprosthetic joint infection: the utility of a simple yet unappreciated enzyme. *J Bone Joint Surg Am.* 93: 2242-2248.

* Rabitzsch G, Mair J, Lechleitner P, Noll F, Hofmann U, Krause EG, Dienstl F, Puschendorf B. (1995). Immunoenzymometric assay of human glycogen phosphorylase BB in diagnosis of ischemic myocardial injury. *Clin Chem* 41:966 -978.

* Radenkovic S, Milosevic Z, Konjevic G, Karadzic K, Rovcanin B, Buta M, Gopcevic K, Jurisic V. (2013). Lactate dehydrogenase, catalase, and superoxide dismutase in tumor tissue of breast cancer patients in respect to mammographic findings. *Cell Biochem Biophys*. 66: 287-295.

* Ram M, Sherer Y, Shoenfeld Y. (2006). Matrix metalloproteinase-9 and autoimmune diseases. *J Clin Immunol*. 26: 299-307.

* Mani R, Sudha Murthy S, Jamil K. (2006). Role of serum lactate dehydrogenase as a bio-marker in therapy related hematological malignancies. *International Journal of Cancer Research*. 2: 383-389.

* Rosalki SB, Roberts R, Katus HA, Giannitsis E, Ladenson JH. (2004). Cardiac biomarkers for detection of myocardial infarction: perspectives from past to present. *Clin Chem*. 50: 2205-2213.

* Schwartz MK. (1992). Enzymes as prognostic markers and therapeutic indicators in patients with cancer. *Clin Chim Acta*. 206: 77-82

* Shen YS, Chen WL, Chang HY, Kuo HY, Chang YC, Chu H. (2012). Diagnostic performance of initial salivary alpha-amylase activity for acute myocardial infarction in patients with acute chest pain. *J Emerg Med*. 43; 553-560.

* Sohar N, Hammer H, Sohar I. (2002). Lysosomal peptidases and glycosidases in rheumatoid arthritis. *Biol Chem*. 383: 865-869.

* Suzuki M, Sai JK, Shimizu T. (2014). Acute pancreatitis in children and adolescents. *World J Gastrointest Pathophysiol*. 5: 416-426.

* Torsteinsdottir I, Hakansson L, Hallgren R, Gudbjornsson, Arvidson NG, Venge P, (1999). Serum lysozyme: a potential marker of monocyte/ macrophage activity in rheumatoid arthritis. *Rheumatology*. 38: 1249-1254.

* Vroon DH, Israili Z. (1990). Alkaline Phosphatase and Gamma Glutamyltransferase, **In**: *Clinical Methods: The History, Physical, and*

Laboratory Examinations (Walker HK, Hall WD, Hurst JW eds) 3rd edition, Boston, Butterworths.

- Wang J, Yuan W, Chen Z, Wu S, Chen J, Ge J, Hou F, Chen Z. (2012). Overexpression of G6PD is associated with poor clinical outcome in gastric cancer. *Tumour Biol.* 33: 95-101.

CHAPTER 6

ENZYMES AS THERAPEUTIC AGENTS

Enzymes are natural catalysts for many biochemical reactions, having high reaction specificity and catalytic efficiency and this has prompted researchers to use enzymes as therapeutic agents for treating metabolic diseases (Shaik, 2014). Therapeutic enzymes that can be used medically either alone or in combination with other treatment help in curing various diseases. These enzymes even in very small amounts act as biological catalysts that can rapidly produce specific effects at physiological pH and temperature. This chapter focuses on the use of enzymes as therapeutic agents in various diseases *viz*, cancer, pancreatic disorders, heart diseases, kidney diseases, respiratory disorders, etc.

Characteristics of Therapeutically Useful Enzymes

Enzymes that degrade selected nutrients or metabolites in circulation should have the following characteristics:

1. High activity and stability at physiological pH.

2. Retention of activity and stability in animal serum and whole blood.

3. High affinity for the substrate (low K_m and high V_{max})

4. Slow clearance from the circulation when injected into animals.

5. No inhibition by its products or other constituents normally found in body fluids.

6. No requirement for exogenous cofactors.

7. Effective irreversibility of the enzymatic reaction under physiological conditions.

8. Availability from a nonpathogenic organism that contains little endotoxin.

Therapeutically useful enzymes are required at a very high degree of purity and specificity. Thus, the sources of such enzymes are chosen with care to avoid any possibility of unwanted contamination by incompatible material and to enable ready purification. Therapeutic enzyme preparations are generally available as lyophilized pure preparations with biocompatible buffering salts and diluent. The costs of such enzymes may be quite high but still comparable to those of competing therapeutic agents or treatments. Enzymes have some unique disadvantages also when used as drugs. For parenteral administration, they must be exhaustively purified to eliminate contaminating toxic materials. Enzymes are generally quite expensive to prepare and they are degraded in the body. They are large molecules with limited distribution within the host. In addition, the enzymes that are foreign proteins to the host are antigenic.

Enzyme therapeutics are broadly divided into three groups for better understanding (Leader et al., 2008).

Table 6.1 Enzyme Therapeutics

GROUP I Replace a protein which is deficient or abnormal	GROUP II Augment an existing pathway	GROUP III Provide a novel function or activity
Pulmonary and Gastrointestinal tract diseases	**Hemostasis and thrombosis**	**Enzymatic degradation of small metabolites**
Pancreatic enzymes	Urokinase	Asparaginase
Metabolic enzyme deficiencies	Streptokinase	Glutaminase

GROUP I **Replace a protein which is deficient or abnormal**	GROUP II **Augment an existing pathway**	GROUP III **Provide a novel function or activity**
Lysosomal hydrolases	Tissue plasminogen activator (TPA)	**Enzymatic degradation of macromolecules**
	Factor IX	Botulinum toxin B
	Factor VIIa	Serratiopeptidase
	Activated protein C	Collagenase
	Thrombin	DNA and RNA enzymes
		Hyaluronidase

GROUP I

Pulmonary and Gastrointestinal Tract Diseases

Pancreatic Enzymes Insufficiency of pancreatic enzymes leads to indigestion and inadequate absorption of fat, protein and carbohydrate, causing steatorrhoea and creatorrhea resulting in abdominal discomfort, weight loss and nutritional deficiencies. This serious condition, known as exocrine pancreatic insufficiency caused by chronic pancreatitis, cystic fibrosis, pancreatic cancer and pancreatic surgery, can be treated by pancreatic enzyme replacement therapy (PERT) (Sikkens et al., 2010). PERT with delayed-release of pancrelipase is now becoming a standard practice since it significantly improves the coefficients of fat and nitrogen absorption as well as clinical symptoms (Nakajima et al., 2012). Currently used pancreatic enzyme supplements contain a mixture of protease, lipase and amylase, known as pancreatin. Bovine pancreatic enzymes appear to be a better alternative compared to porcine and human pancreatic extracts (Löhr et al., 2009). Pancreatic enzymes in the form of enteric-coated mini microspheres are useful in treating patients with Celiac disease (Fieker et al., 2011). Pancreatic enzymes can also be used as supplements in the treatment of indigestion, constipation and bloating.

Recently, Somaraju and Solis-Moya (2014) have evaluated the efficacy and safety of PERT in children and adults with cystic fibrosis. The results show that enteric-coated microspheres score over enteric-coated tablets in terms of stool frequency, abdominal pain and fecal fat excretion. However, it requires more evidence on the long-term effectiveness and risks associated with PERT, relative dosages of enzymes needed for people with different levels of severity of pancreatic insufficiency, optimum time to start treatment and variations based on differences in meals and meal sizes.

Lipases catalyse the hydrolysis of triacylglycerol and phospholipids and they are obtained from bacterial, fungal, plant and animal sources. Lipases have earlier been used in the treatment of gastrointestinal disturbances and dyspepsia. Similase and Vitaline® are high-potency plant enzyme supplements that support digestion of protein, carbohydrates and fats (Hasan et al., 2006). Microbial lipases show significant lipolytic activity and they are stable against proteolytic hydrolysis as well as bile salts resulting in higher lipolytic activity. Oral pancreatic enzyme preparation containing highly stable lipase constitutes a good candidate for enzyme substitution therapy (Domínguez-Muñoz, 2011). The pancrelipase formulations (Creon®, Zenpep® and Pancreaze®) are approved effective treatments for pancreatic enzyme insufficiency (Dhanasekaran and Toskes, 2010). They provide symptomatic relief, prevent morbidity and improve quality of life.

Metabolic Enzyme Deficiencies

Lysosomal Hydrolases   For metabolic enzyme deficiencies, lysosomal hydrolases are good alternatives. Lysosomal hydrolases deficiency causes lysosomal storage diseases (LSDs). It can cause impaired intracellular turnover, degradation and disposal of a variety of substrates present in lysosomes, including sphingolipids, glycosaminoglycans, glycoproteins and glycogen. The pathology of LSDs is typically characterized by the variable association of visceral, ocular, hematological, skeletal and neurological manifestations. A major breakthrough in the treatment of LSDs is the enzyme replacement therapy (ERT) (Parenti, 2009).

ERT refers to a group of glycoprotein enzyme products, each intended to augment or replace the activity of a specific endogenous catabolic enzyme within the cellular lysosomes. All ERT products are administered by intravenous infusion and the infused enzymes are taken up by cells and transported into lysosomes, where they catabolize the specific macromolecule that has accumulated. ERT has been approved for six lysosomal storage diseases (LSDs) which include Gaucher disease, Fabry disease, mucopolysaccharidosis types I, II, and VI and Pompe disease. Some approved enzymes currently used are β-glucocerebrosidase from human placenta (alglucerase) and recombinant enzymes (imiglucérase, Cerezyme®) for the treatment of Gaucher disease. Two currently available preparations for Fabry disease are Replagal(® and Fabrazyme(®. Mucopolysaccharidosis type I (Hurler and Scheie disease), type II (Hunter syndrome) and type VI (Maroteaux-Lamy disease) are treated using laronidase (Aldurazyme®), idursulfase (Elaprase®) and galsulfase (Naglazyme®) respectively (Lachmann, 2011; Parenti et al., 2013). On the other hand, clinical experience with ERT has shown that this approach has limitations, mostly related to the bioavailability of recombinant enzymes and high cost of therapies. To circumvent these problems, different strategies are evaluated to improve delivery of the enzymes to target tissues, including central nervous system.

GROUP II

Hemostasis and Thrombosis

Urokinase Urokinase or urokinase-type plasminogen activator (uPA) is a serine protease which converts plasminogen to plasmin, thus promoting fibrinolysis. uPA, as a thrombolytic agent, is used to treat pulmonary embolism, acute myocardial infarction, severe or massive venous thrombosis, ophthalmic clot and hemorrhage and peripheral arterial occlusion (Crippa, 2007). It is also administered intrapleurally to dissolve fibrinous adhesions thereby improving the drainage of complicated pleural effusions and preventing pleural loculations (Bouros et al., 2007). Urokinase is presently marketed as Kinlytic™ and used as a thrombolytic drug in infarction. It is a life saving, therapeutically important fibrinolytic enzyme used in the treatment of many disorders requiring dissolution of blood clots (Kunamneni et al., 2008). Prolonged thrombolysis with low-dose urokinase could be an alternative approach to therapy in patients with massive pulmonary embolism (Bulpa et al., 2009).

Streptokinase Streptokinase (SK), a bacterial extracellular enzyme, is a single-chain polypeptide that does not have a direct fibrinolytic activity and it converts plasminogen to plasmin. SK is produced by group C β-hemolytic streptococci (trade name - Streptase).SK is used for intravenous administration immediately after the onset of a heart attack in patients, thus reducing the amount of damage to the heart muscle. SK stimulates a cascade system responsible for the production of active plasmin, a proteolytic enzyme which digests fibrin, the main structural component of blood clots. It is usually administered only when the first heart attack occurs (Banerjee et al., 2004). Intrapleural streptokinase is an effective and safe adjunct in facilitating drainage in parapneumonic pleural effusions and pleural empyema (Bouros et al., 2006). SK is used extensively as a thrombolytic agent (Kunamneni et al., 2007). Treatment with antibiotics, tube thoracostomy and intrapleural streptokinase has led to resolution of the pneumonia and empyema without the need for surgical intervention in advanced pregnancy (Nir et al., 2009).

***Tissue Plasminogen Activator* (tPA)** When more than one heart attack occurs, tPA is recommended. The fibrinolytic system comprises an inactive proenzyme, plasminogen which can be converted to the active enzyme plasmin using plasminogen activators and this, in turn, degrades fibrin into soluble fibrin degradation products (Chasman, 2014). The fibrin specificity of tPA has stimulated great interest in its use for thrombolytic therapy. tPA is a poor enzyme in the absence of fibrin, but its activity is dramatically stimulated in the presence of a fibrin cofactor. This fibrin specificity has formed the basis for its development as a thrombolytic agent for treatment of thromboembolic disease. In the therapeutic setting, exogenous administration of tPA is an approved treatment for both myocardial infarction and acute ischemic stroke (Chasman, 2014).

Activase is the first recombinant enzyme drug approved by the USFDA in 1987. Second-generation (recombinant activase) and third-generation plasmin activators (mutants or recombinant reteplase) have several advantages over streptokinase and urokinase viz., i) not antigenic; ii) do not cause immunogenic reactions; iii) absorbed in particular onto fibrin clots and display their effect only there; and iv) no side effects on hemostasis even when introduced into the circulating blood (Maroo and Topol, 2004).

***Factor* IX**   Coagulation Factor IX, a serine protease, is an important protein in the process of hemostasis and normal blood clotting. Hemophilia B is a hereditary bleeding disorder caused by a lack of blood clotting factor IX. Individuals suffering from Hemophilia B are managed with Factor IX (Santagostino et al., 2016) [Trade name: Benefix]. Factor IX is vitamin K dependent and it circulates in the blood as an inactive zymogen. This factor is converted to an active form by factor XIa. The role of this activated factor IX in the blood coagulation cascade is to activate factor X to its active form through interactions with Ca^{2+} ions, membrane phospholipids, and factor VIII. The activation of Factor X then performs a similarly integral step in the blood coagulation cascade.

***Factor* VIIa**   Recombinant FVIIa, NovoSeven® (Novo Nordisk), is the first recombinant product available for the treatment of hemophilia patients having neutralizing antibodies (i.e. inhibitors) towards FVIII or FIX. FVIIa is currently generating significant interest for trauma care and for uncontrolled haemorrhage and coagulopathy. It is approved in the U.S.A. for use in haemophilia patients with inhibitors and in Europe for additional indications such as FVII deficiency and Glanzmann's thrombasthenia.

***Activated Protein* C (Drotrecogin – α)**   It is an antithrombotic which inhibits coagulation factors Va and VII a. This is administered in severe sepsis with a high risk of death. The trade name is Xigris.

Thrombin   Thrombin, a pivotal component of the coagulation cascade, converts fibrinogen into fibrin monomers that multimerize to form stable blood clots. Bovine thrombin, combined with fibrin and sometimes collagen, has been used in various formulations to treat bleeding from surgery or trauma. The bandages containing both bovine thrombin and bovine fibrin have been approved as medical devices to control surface bleeding. In 2008, the FDA has approved topical recombinant human thrombin marketed as Recothrom to prevent small blood vessels from bleeding after surgery. Recombinant human thrombin is found to be as effective as bovine thrombin while being less immunogenic (Bowman et al., 2010).

GROUP III

Enzymatic Degradation of small Metabolites

Asparaginase Asparaginase is an aminohydrolase and it converts asparagine to aspartic acid and ammonia, which leads to cell death. Three types of asparaginase are currently available *viz.*, native asparaginase (trade name – Elspar); pegylated asparaginase (PEG-asparaginase, trade name - Oncospar) derived from *Escherichia coli*; and third one is from *Erwinia chrysanthemi* (Crisantaspase) and it has been found to have the lowest toxicity among a large variety of similar enzymes with known antitumor activity (Rizzari et al., 2013). It is also used as a model enzyme for development of new drug delivery system (Kumar et al., 2011). *Bacillus aryabhattai* ITBHU02 has been reported to be a potential source for the production of L-asparaginase. Actinomycetes, such as *Streptomyces canus*, *S. cyaneus*, *S. exfoliates* and *S. phaeochromogenes* have been shown to be potential producers of glutaminase free L-asparaginase with better therapeutic properties (Velho-Pereira and Kamat 2013; Saxena et al., 2015). It is a potential therapeutic enzyme used in the treatment of acute lymphocytic leukaemia (ALL).

Leukemic cells require a high amount of asparagine for their proliferation and for this they depend on body fluid asparagine. It is well known that tumour cells are deficient in aspartate-ammonia ligase activity, which restricts their ability to synthesise the non-essential amino acid L-asparagine. Hence, they are forced to extract it from body fluids. Asparaginase does not affect the functioning of normal cells which are able to synthesize enough for their own requirements. Administration of L-asparaginase results in depletion of circulating serum asparagine and kills tumor cells. Healthy cells are unaffected as they synthesize asparagine intracellularly with L-asparagine synthetase. The enzyme can be administered intravenously and it is effective only when the asparagine levels within the bloodstream are extremely low (Gurung et al., 2013).

L-asparaginase, in combination with other drugs and radiotherapy, has shown great success in the treatment of ALL. Achievement of complete remission in patients is observed with a few side-effects including pancreatitis, coagulation abnormalities and allergic reactions. Sometimes, tumor cells may develop resistance to L-asparaginase. To overcome this difficulty, the drug is modified by pegylation or immobilization, and also treatment protocols can be modified to increase the efficiency of the drug (Kumar et al., 2014). PEG-asparaginase is widely used for the treatment of children with ALL. It is less immunogenic and has a longer half-life than native *E. coli* asparaginase, which makes it a potent drug with reduced number of doses (Rizzari ct al., 2013).

Glutamine reduction is also necessary for optimal anti-leukemic activity of asparaginase (Saxena et al., 2015). Indeed, both *Escherichia coli* and *Erwinia chrysanthemi* asparaginases possess glutaminase activity also and their administration has shown to reduce serum glutamine level by deamidating glutamine to glutamate and ammonia.

Glutaminase L-Glutaminase is an amidohydrolase that catalyses the conversion of L-glutamine to L-glutamic acid and ammonia. Interest in L-glutaminase has started with the discovery of its antitumor properties (Bauer et al., 1971). L-glutaminase activity is widely distributed in animal, plant tissues and microorganisms including bacteria, fungi and actinomycetes. Although L-glutaminase can be synthesized from both plant and animal sources, microbial source is mostly used for industrial application because of its economic viability, constancy, facility of modulation and optimization ways.

Amino acid depleting enzymes play a significant role in cancer therapy. High rate of glutamine consumption is one of the characteristic features of cancer cells. Glutaminase has been used for the treatment of ALL due to its antileukemic activity. L-glutaminase causes selective death of glutamine dependent tumour cells by blocking the energy source, L-glutamine for cell proliferation (Sarada, 2013).

Kinetic properties of *Streptomyces canarius* L-glutaminase and its anticancer efficiency have been evaluated by Reda (2015). The anticancer activity of L-glutaminase has been tested against five types of cancer cell lines – Hep-G2, MCF7, HCT-116, HeLa and RAW264.7 cells using MTT assay. This study shows that the enzyme has a high efficiency against Hep-G2 and HeLa cell lines. The growth of MCF7 cells is not affected by the treatment, while the HCT-116 and RAW264.7 cells show moderate cytotoxic effects. Results indicate the potential of purified enzyme as a promising candidate for cancer therapy. Dutt et al., (2014) have reported anti-tumour activity of L-glutaminase from *Aspergillus niger*. L-glutaminase produced from *Aspergillus niger* shows anti-tumour property against human breast cancer cell lines MCF- 7 as shown by MTT assay.

Roberts et al., (2001) have patented a therapy for HIV, where L-glutaminase from Pseudomonas sp. 7A is administered to inhibit HIV replication in infected cells. L-glutaminase brings about inhibition of tumour (melanoma) and DNA biosynthesis in affected cells. Cancer cells, especially lymphatic tumor cells, cannot synthesize L-glutamine, and require a large amount of L-glutamine for their rapid growth. The use of L-glutaminase deprives the tumor cells from L-glutamine and causes selective death of L-glutamine dependant tumor cells. L-glutaminase has been used as an efficient anti-retroviral agent in treating acquired immunodeficiency syndrome (AIDS) caused by human immunodeficiency virus (HIV) as it lowers L-glutamine levels in both serum and tissues for prolonged periods. This results in substantial reduction in serum reverse transcriptase activity of the HIV with improved long-term survival benefits (Sarada, 2013). L-glutaminase alone or in combination with asparaginase is mainly used in treating cancer, specifically ALL. It is also a therapeutic agent for retroviral diseases (Balagurunathan et al., 2010). *Achromobacter glutaminasificans* glutaminase–asparaginase is chemically modified by succinylation, and the succinylated enzyme has broader antitumor activity than *E. coli* asparaginase (Elzainy and Ali, 2006).

Enzymatic Degradation of Macromolecules

Botulinum Toxin (BTX) - Type A and B BTX is a neurotoxic protein produced by the bacterium *Clostridium botulinum* and related species. There are eight types of botulinum toxin viz. types A-H. Botulinum toxin types A and B are used in medicine to treat various muscle spasms and diseases characterized by overactive muscle. They are zinc metalloproteases (Craik et al., 2011).

Serratiopeptidase Serratiopeptidase, also known as serralysin/serration protease is a proteolytic enzyme excreted by nonpathogenic Enterobacteria belonging to the Genus Serratia species E 15 isolated from intestine of silkworm *Bombyxmori* L. It has a distinctive ability to digest the dead tissue of the Cocoon of the silkworm and helps the emerging moth. The serratia protease enzyme has a molecular weight of 45,000 to 60,000 and it is a metalloprotease containing a zinc atom which is important for its proteolytic activity. Serratiopeptidase is the most effective agent in reducing inflammation among all enzyme preparations.

In human system, serratiopeptidase breaks down protein deposits such as fibrin. This is used as a natural alternative to steroids and NSAIDs without serious side effects. The enzyme causes proteolysis of all non-vital tissues including blood clots, cysts, tissue plaques and cellular debris and reduces the inflammatory response. Serratiopeptidase binds to $\alpha 2$ – macroglobulin in the blood in 1:1 ratio. It is slowly transferred to the exudates at the site of infection and inflammation. Mechanism of action is by hydrolysis of histamin, bradykinin and serotonin and it indirectly reduces dilatation of blood capillaries and controls permeability. Serratiopeptidase blocks plasmin inhibitors thus helping the fibrinolytic activity of plasmin.

On oral administration, serratiopeptidase is absorbed in GI tract and distributed throughout the body tissues unchanged via systemic circulation. Serratiopeptidase has a few disadvantages as it is susceptible to degradation and vulnerable to acidic pH. The adverse effects may include nausea, vomiting, diarrhoea, epistaxis, haemoptysis and sometimes hypersensitivity. The increased risk of bleeding may also occur when this enzyme is taken with other natural remedies such as garlic, fish oil and turmeric.

It has three mechanisms to reduce inflammation. It breaks down fibrin, the insoluble protein byproducts of blood coagulation and thins the fluids formed from inflammation and injury. It facilitates their drainage which increases the speed of the tissue repair process. It also alleviates pain as it inhibits the release of bradykinin, a specific pain inducing peptide. Serratiopeptidase is prescribed in various specialties such as surgery, orthopaedics, gynaecology and dentistry for its anti-inflammatory and analgesic effects (Bhagat et al., 2013). It is reported to possess anti atherosclerotic effects due to its fibrinolytic and caseinolytic properties and also can be used as a health supplement to prevent cardiovascular morbidity (Jadav et al., 2010). Serratiopeptidase has been found useful in patients suffering from acute or chronic inflammatory disorders of ear, nose or throat such as laryngitis, catarrhal rhino-pharyngitis and sinusitis. It is orally effective equivalent to diclofenac sodium in both acute and chronic phases of inflammation (Bhagat et al., 2013).

Serratiopeptidase is reported to exert a beneficial effect on mucus clearance by reducing neutrophil numbers and altering the viscoelasticity of sputum in patients with chronic airway diseases (Garg et al., 2012). The antibiofilm property of the enzyme may enhance antibiotic efficacy in the treatment of staphylococcal infections (Nakamura et al., 2003). Serratiopeptidase hydrolyses bradykinin, histamine and serotonin responsible for the edematic status, reduces swelling and improves microcirculation and expectoration of sputum (Al-Khateeb and Nusair, 2008).

Collagenase Collagen is a fibrous protein found in skin, tendon, bone, teeth, blood vessels, intestine and cartilage. It corresponds to 30% of the total protein and its main function is structural. There are more than 26 genetically distinct types of collagen, characterized by considerable complexity and diversity in their structure, their splice variants, presence of additional, non-helical domains, their assembly and their function. Each collagen molecule is a small, hard stick formed by inter-lacing in a triple helix of three polypeptide chains called alpha chains.

Collagenases have been used in medical, pharmaceutical, food, cosmetics and textiles segments. In medical applications, it can be used in burns and ulcer treatment, to eliminate scars, for Dupuytren's disease treatment in addition to various types of fibrosis such as liver cirrhosis, to prepare samples for diagnosis and for production of peptides with antioxidant and antimicrobial activities. It plays an extremely important role in the transplant surgery of some specific organs (Wanderley et al., 2017).

Collagenase is a matrix metalloproteinase (MMP) that breaks peptide bonds in collagen. Collagenase was identified for the first time in 1962 (Grossand Lapierre, 1962). True collagenases cleave helical regions of collagen molecules in fibrillar form under various physiological conditions of pH and temperature. Until recently, the production of true collagenases by bacteria has been considered to be confined to only a few species such as Clostridia and a small number of other organisms, notably a strain of *Vibrio alginolyticus*. Unlike animal collagenases that split collagen in its native triple-helical structure, collagenases from bacteria demonstrate broader substrate specificity.

Since collagenase does not harm the cell membrane, it has been greatly used in cell dispersion, tissue separation, and cell culture for many years. For instance, collagenase from *Clostridium histolyticum* has been approved by the Food and Drug Administration (FDA) as a drug (trade name – Santyl) that breaks down the tough cords in Dupuytren and it is widely used in human cell isolation, wound healing, and wound debridement (Cuylits 2013; Coleman et al., 2014; Alipour et al., 2016). Collagenase ointment is more effective than the petrolatum ointment for debridement of necrotic tissue (Ramundo & Gray, 2008). Debridement is the removal of foreign material as well as devitalized or contaminated tissue from a wound bed. This process is applied to the treatment of pressure ulcers, leg ulcers, wounds and burn wounds, and chronic, non-healing, or indolent wounds considered to be stalled in the inflammatory phase of wound healing. This procedure, which is a vital factor in wound bed preparation, can be carried out by surgical, chemical, mechanical, or autolytic removal of the tissue. Collagenase is a safe and an effective choice for debridement of cutaneous lesion and burn wounds (Alipour et al., 2016).

Collagenase helps to break up and remove dead skin and tissue and thus help in repair mechanism. This, in turn, helps antibiotics to work better and speed up body's natural healing process (Ostlie et al., 2012). Collagen-rich matrix constitutes the main barrier to chronic total occlusion (CTO) crossing. Local administration of a human-grade purified collagenase degrades collagen CTO and it effectively facilitates guide-wire crossing in CTO (Strauss et al., 2012). The oxygen–ozone combined collagenase injection for treatment of lumbar disc her niation shows significant reduction in pain and improvement in function and it can be considered as a viable option instead of surgery (Wu et al., 2009). Intravenous injection of type I collagenase digests collagen and can be used as a strategy to improve systemic gene delivery into tumors (Kato et al., 2012). Peyronies disease is an acquired connective-tissue wound-healing disorder of the tunica albuginea of corpus cavernosum and intralesional collagenase has been used for its treatment (Levine, 2013). It has recently been approved by USFDA under the name Xiaflex™ (Auxilium Pharamceuticals, Malvern, PA, USA).

One of the factors in low back pain is the aggregation of collagen tissue and the reduction in the distance between the vertebral spines. Investigations have shown that injection of collagenase can be an improvement in indications of disc herniation (Zhang et al., 2015).

A uterine fibroid is a leiomyoma that is secreted from the flatmuscle layer of the uterus. The disease is often multifarious, and if the uterus contains too many leiomyomata to count, it is mentioned as propagate uterine leiomyomatosis. Having approved by FDA as a drug that does not influence nervous system or blood vessels, *C. histolyticum* collagenase has been evaluated as a suitable treatment for this disease. Injection of *C. histolyticum* collagenase has potential to be used for the treatment of fibroids (Brunengraber et al., 2014).Cellulite or adiposis edematosa is the hernia of subcutaneous fat within fibrous connective tissue that indicates skin dimpling, and the symptoms often are on the pelvic region, lower limbs, and abdomen. Bacterial collagenase is an effective enzyme for the treatment of cellulite (Alipour et al., 2016).

DNA and RNA Enzymes   Ribonucleic acid (RNA) enzymes can be delivered to the cells either endogenously as gene-encoding ribozymes or exogenously as preformed ribozymes. These enzymes can cleave mRNAs and thereby inactivate abnormal gene expression. It is applicable to any disease where a specific gene product can be linked to the initiation and/or perpetuation of the disease. Deoxyribonuclease (DNAse), obtained from *Bacillus* sp. and *Nocardia* sp., degrades deoxyribonucleic acid (DNA) and it has been investigated as a mucolytic agent for the treatment of chronic bronchitis (Sabu et al., 2003). The increased viscosity of pleural pus in patients with pleural empyema is attributed to high concentrations of DNA resulting from the breakdown of phagocytes, bacteria, and other intrapleural cells. Streptodornase, a mixture of four DNAase enzymes, reduces the viscosity of pus by the digestion of DNA; whereas commercial human recombinant DNAse digests DNA and decreases the viscosity of pleural empyema pus without causing allergic reactions known to occur with streptodornase (Simpson et al., 2000). Commercial Pulmozyme® (Dornase α), a DNAse, liquefies accumulated mucus in the lung and also diminishes pulmonary tissue destruction in cystic fibrosis patients (Sabu et al., 2003). Chemically modified DNA enzymes can cleave RNA in the absence of divalent metal ions and such RNA-cleaving DNA enzymes have potential therapeutic applications as antibacterial and antiviral agents (Pradeepkumar and Höbartner, 2012).

The use of DNA enzymes as potent novel drugs for the treatment of inflammatory diseases such as atopic dermatitis has been studied (Schmidts et al., 2012). The two challenges regarding the dermal application of DNA enzymes are the large molecular weight and the sensitivity to DNAses as part of the natural skin flora. To overcome these limitations, they have developed a suitable nano-sized drug carrier system, promising for topical application of DNA enzyme.

Ribozymes are RNA molecules possessing specific catalytic activity and they are capable of catalyzing highly sequence-specific reactions determined by RNA-RNA interactions between the ribozyme and its substrate molecules (Abera et al., 2012). The key to the recognition of and binding to the substrate molecule and the subsequent cleavage reaction resides in the RNA molecule. Ribozymes have been targeted against a myriad of genes, including oncogenes and drug resistance genes, viral diseases such as AIDS, viral hepatitis, mumps virus etc and cellular diseases *viz.*, Alzheimer's disease, cancer, diabetes, rheumatoid arthritis etc (Puerta-Fernández et al., 2003).

RNA-based therapeutics are investigated for diseases ranging from genetic disorders to HIV infection to various cancers. The emerging drugs include therapeutic ribozymes, aptamers, and small interfering RNAs (siRNAs), demonstrating the unprecedented versatility of RNA. Hammerhead ribozymes are catalytic RNA molecules capable of inducing the site-specific cleavage of a phosphodiester bond within an RNA molecule (Grassi et al., 2004). Thus, they can be used to reduce the intracellular level of a specific mRNA coding for a protein which affects cellular metabolism or environment, causing disease. Nucleic acids are potentially immunogenic and typically require a delivery tool to be utilized as therapeutics. Hence, improved synthetic delivery carriers and chemical modifications of the RNA therapeutics are necessary (Grassi et al., 2004; Burnett and Rossi, 2012). Hepatitis C virus genome is present exclusively in RNA form during replication. Various nucleic acid-based therapeutic approaches targeting the hepatitis C virus genome, such as ribozymes, aptamers, siRNAs, and antisense oligonucleotide, have been suggested as potential tools against hepatitis C virus (Lee et al., 2013).

Hyaluronidase Hyaluronidases are glycosidases with both hydrolytic and transglycosidase activities and they catalyze the hydrolysis of hyaluronic acid, lower the viscosity and increase tissue permeability (El-Safory et al., 2010).In ophthalmology, hyaluronidase is most often used as an adjunct to local anaesthesia for retrobulbular, peribulbular and sub-Tenon's block. It decreases intraocular pressure, reduces distortion of the surgical site, decreases incidence of post operative strabismus and limited local anaesthetic myotoxicity (Silverstein et al., 2012). Hyaluronidase (trade name – Amphadase) liquefies the vitreous haemorrhage as demonstrated in a phase III trial in diabetic patients and it can be used as an alternative or adjunct to conventional mechanical vitrectomy (Gandorfer, 2008). Hyaluronidase is used therapeutically in combination with other drugs to speed up delivery and absorption and to diminish discomfort due to subcutaneous or intramuscular injection of fluid, to increase the effectiveness of local anaesthesia and as a spreading factor to improve better penetration of chemotherapeutic drug into tumors. Hyaluronidase treatment destroys the hyaluronate coat surrounding tumor cells and allows lymphocytes to approach the tumor membrane to enhance the cytotoxic action of immune response (El-Safory et al., 2010).

Further, hyaluronidase disrupts intercellular adhesion and chemosensitizes tumor cells by a mechanism, independent of increased drug penetration in cancer chemotherapy. Hyaluronic acid levels are elevated in several cancers and its degradation using hyaluronidase has been shown to enhance the action of various chemotherapeutic agents in patients. Hyaluronidase facilitates penetration and decreases interstitial fluid pressure, permitting anticancer agents to reach malignant cells. Moreover, it has been proposed that hyaluronidase may itself have intrinsic anticancer activity (Guedan et al., or et al., 2010).

A synopsis of enzyme therapy is given in Table 6.2.

Table 6.2   Therapeutic applications of enzymes

S.No	Therapeutic enzymes	Diseases treated
1	Pancreatic enzymes	Exocrine pancreatic insufficiency (Sikkens et al., 2010; Nakajima et al., 2012); fat malabsorption (Lohr et al., 2009); Exocrine celiac and Crohn's disease (Feiker et al., 2011)
2	Lysosomal hydrolases	Metabolic enzyme deficiencies (Parenti, 2009)
3	Urokinase	Thrombolytic agent, Treatment of pulmonary embolism, acute myocardial infarction, ophthalmic clot & hemorrhage empyema (Crippa 2007; Kunamneni et al., 2008; Bulpa et al., 2009)
4	Streptokinase	Thrombolytic agent (Kunamneni et al., 2007); Treatment of heart attack, empyema & pneumonia (Bouros et al., 2006; Nir et al., 2009)

S.No	Therapeutic enzymes	Diseases treated
5	Tissue plasminogen activator	Myocardial infarction and acute ischemic stroke (Chasman, 2014).
6	Factor IX	Treatment of Hemophilia B (Santagostino et al., 2016)
7	Factor VII a	Treatment of Hemophilia patients having neutralizing antibodies; trauma care; uncontrolled hemorrhage and coagulopathy (Baker et al., 2005)
8	Activated Protein C	Severe sepsis with a high risk of death (Baker et al., 2005)
9	Thrombin	Treatment of bleeding from surgery or trauma (Bowman et al., 2010).
10	Asparaginase	Treatment of tumour and lymphoproliferative disorders (Kumar et al., 2014; Rizzari et al., 2013)
11	L-glutaminase	Treatment of cancer & HIV therapy (Sarada, 2013); antitumour activity (Balagurunathan et al., 2010)
12	Botulinum toxin Type A and Type B	Treatment of various muscle spasms (Craik et al., 2011)
13	Serratiopeptidase	Treatment of pain & inflammatory disorders, laryngitis & sinusitis; (Jadav et al., 2010; Bhagat et al., 2013)

S.No	Therapeutic enzymes	Diseases treated
14	Collagenase	Wound debridement. Treatment of skin ulcers and burns (Ramundo and Gray, 2008); Dupuytren's contractures (Cuylits, 2013); facilitation of guide-wire crossing in CTO (Strauss et al., 2012); tumour (Kato et al., 2012)
15	DNA & RNA Enzymes	Antibacterial and antiviral diseases (Pradeepkumar and Hobartner, 2012); viral & cellular diseases (Puerta-Fernandez et al., 2003)
16	Hyaluronidase	Spreading factor (El-Safory et al., 2010); Antitumour activity (El-Safory et al., 2010); Fasciotomy & Virectomy (Gandorfer, 2008)

The large molecular size of enzymes limits their distribution in the body after parenteral administration, but allows for local therapeutic effects. These properties have led to numerous applications, first as crude topical and oral preparations with proteolytic and hydrolytic activity, and later as highly purified enzymes for the treatment of cancer, clotting disorders, genetic defects, inflammation, digestive problems, drug toxicities, and kidney failure. Native enzymes are often unsuitable for these therapeutic uses because of their antigenicity, rapid inactivation, and limited distribution in the body. Recently, many techniques have been developed to improve the therapeutic properties of enzymes. These include (i) soluble chemical modifications, (ii) insoluble chemical modifications or binding to surfaces and (iii) encapsulation in biodegradable or inert materials.

References

- Abera G, Berhanu G, Tekewe A. (2012). Ribozymes: nucleic acid enzymes with potential pharmaceutical applications—a review. *Pharmacophore.* 3: 164 -178.

- Alipour H, Raz A, Zakeri S, Djadid ND. (2016). Therapeutic applications of collagenase (metalloproteases): A review. *Asian Pac J Trop Biomed.* 6: 975–981.

- Al-Khateeb TH, Nusair Y. (2008). Effect of the proteolytic enzyme serrapeptase on swelling, pain and trismus after surgical extraction of mandibular third molars. *Int J Oral Maxillofac Surg.* 37: 264-268.

- Baker SS, Borowitz D, Duffy L, Fitzpatrick L, Gyamfi J, Baker RD. (2005). Pancreatic enzyme therapy and clinical outcomes in patients with cystic fibrosis. *J Pediatr.* 146: 189–193.

- Balagurunathan R, Radhakrishnan M, Somasundaram S T. (2010). L-glutaminase producing actinomycetes from marine sediments–selective isolation, semi quantitative assay and characterization of potential strain. *Aust J Basic Appl Sci.* 4: 698 – 705.

- Banerjee A, Chisti Y, Banerjee UC. (2004). Streptokinase—a clinically useful thrombolytic agent. *Biotechnol Adv.* 22: 287-307.

- Bauer K, Bierling R, Kaufman W. (1971). Effect of L-glutaminase from *Pseudomonas aureofaciens* in experimental tumors. *Naturwissenschaften.* 58: 526–527.

- Bhagat S, Agarwal M, Roy V. (2013). Serratiopeptidase: a systematic review of the existing evidence. *Int J Surg.* 11: 209-217.

- Bouros D, Antoniou K, Light RW. (2006). Intrapleural streptokinase for pleural infection. *BMJ.* 332: 133-134.

- Bouros D, Tzouvelekis A, Antoniou KM, Heffner JE. (2007). Intrapleural fibrinolytic therapy for pleural infection. *Pulm Pharmacol Ther.* 20: 616-626.

- Bowman LJ, Anderson CD, Chapman WC. (2010). Topical recombinant human thrombin in surgical haemostasis. *Semin Thromb Hemostasis.* 36: 477–484.

- Brunengraber LN, Jayes FL, Leppert PC. (2014). Injectable *Clostridium histolyticum* collagenase as a potential treatment for uterine fibroids. *Reprod Sci.* 21: 1452-1459.

- Bulpa P, Carbutti G, Osselaer JC, Lawson G, Dive A, Evrard P. (2009). Low-dose urokinase in massive pulmonary embolism when standard thrombolysis is contraindicated. *Chest.* 136: 1141-1143.

- Burnett JC, Rossi JJ. (2012). RNA-based therapeutics: current progress and future prospects. *Chem Biol.* 19: 60-71.

- Chasman DI. (2014). New pathway for tissue-type plasminogen activator regulation. *Arterioscler Thromb Vasc Biol.* 34: 964-965.

- Coleman S, Gilpin D, Kaplan FT, Houston A, Kaufman GJ,Cohen BM, Jones N, Tursi JP. (2014). Efficacy and safety of concurrent collagenase *Clostridium histolyticum* injections for multiple Dupuytren contractures. *J Hand Surg Am.* 39: 57-64.

- Craik CS, Page MJ, Madison EL. (2011). Proteases as therapeutics. *Biochem J.* 435: 1–16.

- Crippa MP. (2007). Molecules in focus: Urokinase-type plasminogen activator. *Int J Biochem Cell Biol.* 39:690-694.

- Cuylits N. (2013). News in the treatment of Dupuytren's disease: from surgery to collagenase's injection. *Rev Med Brux.* 34:283-286.

- Dhanasekaran R, Toskes PP. (2010). Pancrelipase for pancreatic disorders: An update. *Drugs Today* (Barc). 46:855-866.

- Domínguez-Muñoz JE. (2011). Pancreatic enzyme therapy for pancreatic exocrine insufficiency. *Gastroenterol Hepatol.* 7: 401-403.

- Dutt SPLNSN. Siddalingeshwara KG, Karthic J, Pramod T, Vishwantha T. (2014). Antitumour property of L-glutaminase from *Aspergillus oryzae* through submerged fermentation. *Int J Curr Microbiol Appl Sci.* 3: 819–823.

• El-Safory NS, Fazary AE, Lee CK. (2010). Hyaluronidases, a group of glycosidases: Current and future perspectives. *Carbohyd Polym.* 81:165-181.

• Elzainy TA, Ali TH. (2006). Detection of antitumor glutaminase - asparaginase in the filamentous fungi. *J Appl Sci.* 6:1381-1395.

• Fieker A, Philpott J, Armand M. (2011). Enzyme replacement therapy for pancreatic insufficiency: present and future. *Clin Exp Gastroenterol.* 4:55-73.

• Gandorfer A. (2008). Enzymatic vitreous disruption. Eye (Lond). 22: 1273-1277.

• Garg R, Aslam S, Garg A, Walia R. (2012). A prospective comparative study of serratiopeptidase and aceclofenac in upper and lower limb soft tissue trauma cases. *Int J Pharmacol Pharmaceut Technol.* 1: 11-16.

• Grassi G, Dawson P, Guarnieri G, Kandolf R, Grassi M. (2004). Therapeutic potential of hammerhead ribozymes in the treatment of hyper-proliferative diseases. *Curr Pharm Biotechnol.* 5: 369-386.

• Gross J, Lapiere CM. (1962). Collagenolytic activity in amphibian tissues: a tissue culture assay. *Proc Natl Acad Sci USA.* 48: 1014-1022.

• Guedan S, Rojas JJ, Gros A, Mercade E, Cascallo M, Alemany R. (2010). Hyaluronidase expression by an oncolytic adenovirus enhances its intratumoral spread and suppresses tumor growth. *Mol Ther.* 18: 1275-1283.

• Gurung N, Ray S, Bose S, Rai V. (2013). A broader view: microbial enzymes and their relevance in industries, medicine, and beyond. *Bio Med Research International.* 2013: 329121.

• Hasan F, Shah AA, Hameed A. (2006). Industrial applications of microbial lipases. *Enzyme Microb Tech.* 39: 235-251.

• Jadav SP, Patel NH, Shah TG, Gajera MV, Trivedi HR, Shah BK. (2010). Comparison of anti-inflammatory activity of serratiopeptidase and diclofenac in albino rats. *J Pharmacol Pharmacother.* 1: 116-117.

- Kato M, Hattori Y, Kubo M, Maitani Y. (2012). Collagenase-1 injection improved tumor distribution and gene expression of cationic lipoplex. *Int J Pharmaceut.* 423: 428-434.

- Kumar K, Kaur J, Walia S, Pathak T, Aggarwal D. (2014). L-asparaginase: an effective agent in the treatment of acute lymphoblastic leukemia. *Leuk Lymphoma.* 55: 256-262.

- Kumar S, Venkata Dasu V, Pakshirajan K. (2011). Studies on pH and thermal stability of novel purified L-asparaginase from *Pectobacterium carotovorum* MTCC 1428. *Microbiology,* 80: 355-362.

- Kunamneni A, Abdelghani TT, Ellaiah P. (2007). Streptokinase--the drug of choice for thrombolytic therapy. *J Thromb.* 23: 9-23.

- Kunamneni A, Ravuri BD, Ellaiah P, Prabhakhar T, Saisha V. (2008). Urokinase-a strong plasminogen activator. *Biotechnol Mol Biol Rev.* 3: 58-70.

- Lachmann RH. (2011). Enzyme replacement therapy for lysosomal storage diseases. *Curr Opin Pediatr.* 23: 588-593.

- Leader B, Baca QJ, Golan DE. (2008). Protein therapeutics: a summary and pharmacological classification. *Nat Rev Drug Discov.* 7: 21-39.

- Lee CH, Kim JH, Lee SW. (2013). Prospects for nucleic acid-based therapeutics against hepatitis C virus. *World J Gastroenterol.* 19: 8949-8962.

- Levine LA. (2013). Peyronie's disease: A contemporary review of non-surgical treatment. *Arab J Urol,* 11: 278-283.

- Löhr JM, Hummel FM, Pirilis KT, Steinkamp G, Korner A, Henniges F. (2009). Properties of different pancreatin preparations used in pancreatic exocrine insufficiency. *Eur J Gastroenterol Hepatol* 21: 1024-1031.

- Maroo A, Topol EJ. (2004). Clot busters!!-discovery of thrombolytic therapy for heart attack & stroke. *J Thromb Haem.* 2: 1-4.

- Nakajima K, Oshida H, Muneyuki T, Kakei M. (2012). Pancrelipase: an evidence-based review of its use for treating pancreatic exocrine insufficiency. *Core Evid.* 7: 77-91.

• Nakamura S, Hashimoto Y, Mikami M, Yamanaka E, Soma T, Hino M, Azuma A, Kudoh S. (2003). Effect of the proteolytic enzyme serrapeptase in patients with chronic airway disease. *Respirology*. 8: 316-320.

• Nir S, Gadi L, Mony S, Meir M, Jaime G, Amir E, Uzi I, Gil L, Yaron I, Neville B. (2009). Successful use of streptokinase for the treatment of empyema thoracis during advanced pregnancy: A case report. *Respiratory Med CME*. 2: 21-24.

• Ostlie DJ, Juang D, Aguayo P, Pettiford-Cunningham JP, Erkmann EA, Rash DE, Sharp SW, Sharp RJ, St Peter SD. (2012). Topical silver sulfadiazine vs collagenase ointment for the treatment of partial thickness burns in children: a prospective randomized trial. *J Pediatr Surg*. 47: 1204-1207.

• Parenti G, Pignata C, Vajro P, Salerno M. (2013). New strategies for the treatment of lysosomal storage diseases. *Int J Mol Med*. 31: 11-20.

• Parenti G. (2009). Treating lysosomal storage diseases with pharmacological chaperones: from concept to clinics. *EMBO Mol Med*. 1: 268-279.

• Pradeepkumar PI, Höbartner C. (2012). RNA-Cleaving DNA enzymes and their potential therapeutic applications as antibacterial and antiviral agents, In: *From Nucleic Acids Sequences to Molecular Medicine RNA Technologies*, (Eds. VA Erdmann & J Barciszewski, Springer, Berlin Heidelberg) 371.

• Puerta-Fernández E, Romero-López C, Barroso-delJesus A, Berzal-Herranz A. (2003). Ribozymes: recent advances in the development of RNA tools. *FEMS Microbiol Rev*. 27: 75-97.

• Ramundo J, Gray M. (2008). Enzymatic wound debridement. *J Wound Ostomy Continence Nurs*. 35: 273-280.

• Reda FM. (2015). Kinetic properties of *Streptomyces canarius* L-glutaminase and its anticancer efficiency. *Braz J Microbiol*. 46: 957–968.

• Rizzari C, Conter V, Starý J, Colombini A, Moericke A, Schrappe M. (2013). Optimizing asparaginase therapy for acute lymphoblastic leukemia. *Curr Opin Oncol,* 25 (Suppl - 1):S1-9.

- Roberts J, MacAllister TW, Sethuraman N, Freeman AG. (2001). Genetically engineered glutaminase and its use in antiviral and anticancer therapy, US patent. 6: 6312939.

- Sabu A. (2003). Sources, Properties and Applications of Microbial Therapeutic Enzymes. *Ind J Biotechnol.* 2: 334-341.

- Santagostino E, Martinowitz U, Lissitchkov T, Pan-Petesch B, Hanabusa H, Oldenburg J, Boggio L, Negrier C, Pabinger I, von Depka Prondzinski M, Altisent C, Castaman G, Yamamoto K, Álvarez-Roman MT, Voigt C, Blackman N, Jacobs I, PROLONG-9FP Investigators Study Group. (2016). Long-acting recombinant coagulation factor IX albumin fusion protein (rIX-FP) in hemophilia B: results of a phase 3 trial. *Blood* 127: 1761-1769

- Sarada KV. (2013). Production and applications of L-Glutaminase using fermentation technology. *Asia Pac J Res.* 1: 1-4.

- Saxena A, Upadhyay R, Kango N. (2015). Isolation and identification of actinomycetes for production of novel extracellular glutaminase free L-asparaginase. *Indian J Exp Biol.* 53: 786-793.

- Schmidts T, Marquardt K, Schlupp P, Dobler D, Heinz F, Mäder U, Garn H, Renz H, Zeitvogel J, Werfel T, Runkel F. (2012). Development of drug delivery systems for the dermal application of therapeutic DNAzymes. *Int J Pharm.* 431: 61-69.

- Shaik, M.M. 2014. Enzymes as Therapeutic Agents in Alzheimer's Disease. *J. Biomol. Res. Ther*, 3:e129.

- Sikkens ECM, Cahen DL, Kuipers EJ, Bruno MJ. (2010). Pancreatic enzyme replacement therapy in chronic pancreatitis. *Best Pract Res Clin Gastroenterol*, 24: 337-347.

- Silverstein SM, Greenbaum S, Stern R. (2012). Hyaluronidase in ophthalmology. *J Appl Res.* 12: 1-13.

- Simpson G, Roomes D, Heron M. (2000). Effects of streptokinase and deoxyribonuclease on viscosity of human surgical and empyema pus. *Chest.* 117: 1728-1733.

- Somaraju UR, Solis-Moya A. (2014). Pancreatic enzyme replacement therapy for people with cystic fibrosis. *Cochrane Database Syst Rev.* 10: CD008227

- Strauss BH, Osherov AB, Radhakrishnan S, Mancini GB, Manners A, Sparkes JD, Chisholm RJ. (2012). Collagenase Total Occlusion-1 (CTO-1) trial: a phase I, dose-escalation, safety study. *Circulation.* 125: 522-528.

- Velho-Pereira S, Kamat NM. (2013). Actinobacteriological research in India. *Indian J Exp Biol.* 51: 573-596.

- Wanderley MCA, JMWD Netoa, JLL Filhoa, CA Limab, JAC Teixeirac, ALF Portod. (2017). Collagenolytic enzymes produced by fungi: a systematic review. *Brazilian Journal of Microbiology.* 48: 13-24.

- Wu Z, Wei LX, Li J, Wang Y, Ni DH, Yang P, Zhang Y. (2009). Percutaneous treatment of non-contained lumbar disc herniation by injection of oxygen–ozone combined with collagenase. *Eur J Radiol,* 72: 499-504.

- Zhang D, Zhang Y, Wang Z, Zhang X, Sheng M. (2015). Target radiofrequency combined with collagenase chemonucleolysis in the treatment of lumbar intervertebral disc herniation. *Int J Clin Exp Med.* 8: 526-532.

CHAPTER 7

WHAT IS AHEAD?

Biomarker studies have entered a whole new era and hold promise for early diagnosis and effective treatment of many diseases (Biomarkers Definitions Working Group, 2001). Studies have shown that a panel of biomarkers could cumulatively present a higher sensitivity and specificity than any single biomarker for disease diagnosis.

DIAGNOSTIC PROTEINS FOR FUTURE

Glycated Albumin (GA)

Glycated hemoglobin (HbA1c) is not recommended in clinical situations such as hemolytic, secondary or iron deficiency anemia, hemoglobinopathies, pregnancy and uremia which may interfere with the metabolism of hemoglobin. GA is a test that reflects short-term glycemia and is not influenced by situations that falsely alter A1c levels. GA is the higher glycated portion of fructosamine and it is measured by a standardized enzymatic methodology, easy and fast to perform. These characteristics have shown the importance of GA as a marker for monitoring and screening for diabetes mellitus (DM) as well as a predictor of long-term outcome of the disease (Freitas et al., 2017).

Angiopoietin 2

Angiopoietin 2 is an angiogenic factor, which exerts a selective effect on endothelium by binding with Tie-2 receptor. Under physiological conditions, it is stored in Weibel-Palade bodies. Its release in the systemic circulation is triggered by factors viz., hypoxia, inflammation or mechanical injury. Concentrations of this factor reflect the extent of endothelial activation (El-Banawy et al., 2012). Sporek et al., (2016) have reported that Angiopoietin 2 serves as a predictor of acute pancreatic-renal syndrome in patients with acute pancreatitis.

Podocalyxin and Nephrin

Podocalyxin and nephrin are suggested as urinary markers of podocyte dysfunction in glomerular diseases. Podocytes are cells in the Bowman's capsule in the kidneys that wrap around capillaries of the glomerulus. Podocalyxin and nephrin are podocyte – associated proteins and they play key roles in maintaining the structural and functional integrity of the kidney's filtration barrier. While podocalyxin offers the structural support to the capillary loop, nephrin impedes the filtration of large molecules into the urinary space checking proteinuria. Filtration of blood and removal of wastes is the main glomerular function accomplished with the aid of podocytes and their associated proteins.

Podocytes are injured in a number of glomerular diseases including diabetic nephropathy, membranous nephropathy, focal segmental glomerulosclerosis and lupus nephritis. The detectability of podocalyxin and nephrin in glomerular diseases is getting established and they are potential noninvasive diagnostic biomarkers (Akanwasa et al., 2018). The levels of both podocalyxin and nephrin are significantly increased, reflecting podocyte injury in the early stages of glomerular diseases. Future research should aim to elucidate the diagnostic and prognostic capabilities of podocalyxin and nephrin.

Osteoprotegrin (OPG) and Receptor Activator of Nuclear Factor Kappa B Ligand (RANKL)

Osteoclasts are specialized multinucleated giant cells that develop from monocyte-hematopoietic cells. Harada and Rodan (2003) have specified OPG, cathepsin K and chloride channel 7 (ClCN7) as rate-limiting agents for osteoclast differentiation and function. OPG blocks the transcription factor (TF) of receptor activator of nuclear activator kappa B (RANK) and RANKL docking. RANK and RANKL are key proteins regulating osteoclast function (Masella and Meister, 2006).

In periodontal disease, the ratio of OPG to RANKL is altered to favour RANKL and this may contribute to the alveolar bone resorption seen in periodontitis (Crotti et al., 2003). Therapies to promote the OPG/RANKL ratio in favour of OPG - ie., the normal state in the periodontal ligament - may well reduce the alveolar bone resorption of the disease. Thus, OPG and RANKL could be developed as prognostic markers for periodontitis and in appliance activation.

THERAPEUTIC PROTEINS FOR FUTURE

Although therapeutic proteins have become prominent in drug discovery and development, there are certain challenges still to be addressed (Akash et al., 2015). The most important feature of these therapeutic proteins is their safety and clinical efficacy which can be achieved by a combination of several factors, including disease state, target biology, potency, safety margin, dosing and selection of patient population (Strohl, 2009). The safety and clinical efficacy of the administered therapeutic substance can differentiate the best therapeutic agent from the others.

Immunogenicity has become a major challenge for drug efficacy and disease management. Even a small amount of particulate material in a therapeutic protein formulation is considered to cause immunogenicity. It is ideal to assess the potential of immunogenicity during the discovery phase itself and ultimately select a molecule with a minimal immunogenic profile as a clinical therapeutic candidate. Immunogenicity and instability of therapeutic proteins can be nullified by modifying the protein structure via alteration of amino acid sequences and/or optimizing the formulation of therapeutic proteins using different types of polymers. The polymers, used for the incorporation of therapeutic proteins, must be inert, biocompatible and most preferentially biodegradable in nature.

Hepatic metabolism and rapid elimination is a major problem in the clinical application of therapeutic proteins. To overcome this, efforts are made to include the noninvasive delivery of therapeutic proteins through routes (other than the oral route) that bypass the hepatic metabolism of the administered therapeutic proteins. In addition, majority of therapeutic proteins have a short biological half-life. This problem could be overcome by encapsulating and/or conjugating the desired therapeutic proteins with biocompatible polymers and using fusion protein technology to extend the half-life of therapeutic proteins (Akash et al., 2015).

DIAGNOSTIC ENZYMES FOR FUTURE

***Thymidine Kinase 1* (TK1):**   TK1 is an enzyme involved in nucleic acid synthesis and is therefore co nsidered to be a marker in tumor proliferation. TK1 is almost undetectable in normal serum, but increases to varying degrees in malignant tumors, depending on their type, stage, growth rate and presence or absence of treatment. It may be more useful to combine TK1 with other tumor markers for the diagnosis and monitoring of the tumor therapy outcome (Xiang et al., 2013).

THERAPEUTIC ENZYMES FOR FUTURE

Psychrophilic Proteases

The proteases so far approved by the US FDA are sourced from a range of mammals or bacteria that exist or have adapted to moderate temperatures i.e., mesophilic organisms. However, proteases, derived from organisms from cold environments named as cold adapted or psychrophilic proteases, generally have high specific activity, low substrate affinity and high catalytic rates at low and moderate temperatures. Cold-adapted proteases have shown promise as therapeutic modalities for cosmeceutical applications (by reducing glabellar [frown] lines) and in a number of disease conditions, including bacterial infections (by disrupting biofilms to prevent bacterial infection), topical wound management (when used as a debridement agent to remove necrotic tissue and fibrin clots), oral/dental health management (by removing plaque and preventing periodontal disease), and in viral infections (by reducing the infectivity of viruses, such as human rhinovirus 16 and herpes simplex virus) (Fornbacke and Clarsund, 2013).

Antioxidant Enzymes

Antioxidant enzymes such as superoxide dismutase, catalase, heme oxygenase, glutathione peroxidase and peroxiredoxin are the scavengers of reactive oxygen species (ROS) and they are used as indices/therapeutic agents for treatment of oxidative stress related disorders including neurological and neuroinflammatory diseases. Heme oxygenases have both antioxidative and antiinflammatory properties whereas peroxiredoxins play an important role in removing hydrogen peroxide (Schreibelt et al., 2007).

Amyloid β-protein (Aβ) cleaving Proteases

Aβ is the main component of amyloid plaques, which accumulate abnormally in the brains of patients with Alzheimer's disease. Leissring (2008) has detailed the involvement of specific Aβ cleaving proteases in the etiology and potential treatment of Alzheimer's disease. The Aβ-cleaving proteases are referred as novel class of enzymes that may serve as therapeutic agents.

L-lysine α-oxidase

L-lysine α-oxidase catalyzes mainly oxidative deamination of L-lysine resulting in a decreased level of the essential amino acid L-lysine and producing α-keto-ε-aminocaproic acid and hydrogen peroxide. This possibly provides the basis for the unique properties of L-lysine α-oxidase viz., cytotoxic, antitumor, antimetastatic, anti-invasive, antibacterial, and antiviral activities, as well as immunomodulating effect (Joseph and Rajan, 2011).

Chondroitinase ABC (ChABC)

Chondroitin sulphate proteoglycans (CSPGs) are potent inhibitors of growth in the adult CNS and use of ChABC reduces the CSPG inhibition of spinal cord injury and promotes regeneration of injured axons, plasticity of uninjured pathways and neuroprotection of injured projection neurons (Bradbury and Carter, 2011).

Alginate Lyase

Alginate lyases are promising therapeutic candidates for treating mucoid *Pseudomonas aeruginosa* infections. The mucoid *P. aeruginosa* strains frequently isolated from cystic fibrosis patients have alginate exopolysaccharide which acts as a barrier against host's immune defense and antibiotic treatment. It is reported that alginate lyase capable of degrading alginate polymer can be used in combination with antibiotics for the treatment of cystic fibrosis (Lamppa and Griswold, 2013).

Hemocoagulase

Hemocoagulase is an enzyme complex, isolated from snake venom, possessing coagulative and antihemorrhagic property. It accelerates the conversion of fibrinogen to fibrin polymer and promotes the interaction of platelets with fibrin clot. The clot thus formed is resistant to plasmin (Aslam et al., 2013; Shenoy et al., 2014). Hemocoagulase is suggested to play a good hemostatic role in the hemorrhagic capillary in abdominal incision, in cases of cleft palate and septum deviation during plastic surgery, and in the control of intraocular bleeding during vitreous surgery. Wei et al., (2010) have conducted phase III clinical trial to evaluate the effect of hemocoagulase Agkistrodon obtained from chinese moccasin snake venom.

Bromelain

Bromelain is a crude, aqueous extract obtained from the pineapple (*Ananas comosus* Merr.) containing a mixture of proteolytic enzymes, referred to as sulfhydryl proteases. Pillai et al., (2013) have demonstrated that bromelain has the potential of being used as an effective anticancer agent for malignant peritoneal mesothelioma. During the treatment of patients with burn wounds, surgical escharotomy may cause considerable blood loss. Debridase, a bromelain derived enzymatic preparation, is capable of lysing the burn eschar within 4 h, obviating the need for surgical debridement. It has an affinity to burned necrotic tissue and does not damage healthy skin (Krieger et al., 2005).

Lipase

Another futuristic development is the restoration of a patient's own bio-engineered lipase production with gene therapy for treating lipoprotein lipase deficiency (Gaudet et al., 2012).

Methionine Gamma-Lyase (MGL)

MGL degrades sulfur-containing amino acids to alpha-keto acids, ammonia and thiols. Because sulfur-containing amino acids play a role in multiple biological processes, the regulation of these amino acids is essential. Some tumors, such as glioblastomas, medullo blastoma and neuroblastoma are much more sensitive to methionine starvation than the normal tissues. Therefore, methionine depletion arises as a relevant therapeutical approach to treat cancer. For that reason, MGL has been studied to decrease the methionine levels in blood serum and decrease the tumor growth and also to kill, by starvation, those malignant cells (Fernandes et al., 2017).

Therefore, a better insight into the routes of administration for protein therapeutics and the drug absorption mechanisms is necessary and studies have to be focused on these aspects (Akash et al., 2015). With a large number of protein therapeutics presently in use, it could be visualized that protein and enzyme therapeutics will have a significant role in medicine in the years to come.

References

- Akanwasa G, Jienhua L, Guixue C, Changjuan A, Xiaosong Q. (2018). Urine markers of podocyte dysfunction: a review of podocalyxin and nephrin in selected glomerular diseases. *Biomark Med.* 12: 927-935.

- Akash MSH, Rehman K, Tariq M and Chan S. (2015). Development of therapeutic proteins: advances and challenges. *Turkish J Biol.* 39: 1-16.

- Aslam S, Francis PG, Rao BHS, Ummar M, Issac JK, Nair RB. (2013). A double blind study on the efficacy of local application of hemocoagulase solution in wound healing. *J Contemp Dent Pract.* 14: 394-400.

- Biomarkers Definitions Working Group. (2001). Biomarkers and surrogate end points: preferred definitions and conceptual framework. *Clin Pharmacol Ther.* 69: 89-95.

- Bradbury EJ, Carter LM. (2011). Manipulating the glial scar: chondroitinase ABC as a therapy for spinal cord injury. *Brain Res Bull.* 10: 306-316.

- Crotti T, Smith MD, Hirsch R, Soukoulis S, Weedon H, Capone M, Ahern MJ, Haynes D. (2003). Receptor activator NF Kappa B ligand (RANKL) and OPG protein expression in periodontitis. *J Periodontal Res.* 38: 380-387.

- El-Banawy HS, Gaber E W, Maharem D A, Matrawy K A. (2012). Angiopoietin-2, endothelial dysfunction and renal involvement in patients with systemic lupus erythematosus. *J Nephrol.* 25: 541–550.

- Fernandes HS, Silva Teixeira CS, Fernandes PA, Ramos MJ, Cerqueira NM. (2017). Amino acid deprivation using enzymes as a targeted therapy for cancer and viral infections. *Expert Opin Ther Pat.* 27: 283-297.

- Fornbacke M, Clarsund M. (2013). Cold-Adapted Proteases as an Emerging Class of Therapeutics. *Infect Dis Ther.* 2: 15–26.

- Freitas PAC, Enlert LR, Comargo JL. (2017). Glycated albumin: a potential biomarker in diabetes. *Arch Endocrinol Metab.* 61: 296-304.

- Gaudet D, Methot J, Kastelein J. (2012). Gene therapy for lipoprotein lipase deficiency. *Curr Opin Lipidol.* 23: 310-320.

- Harada S, Rodan GA. (2003). Control of osteoblast function and regulation of bone mass. *Nature.* 423: 349-354.

- Joseph B, Rajan SS. (2011). L-lysine α oxidases from fungi as an anti-tumour agent. *Adv Biotech.* 10: 27-30.

- Krieger Y, Rosenberg L, Lapid O, Glesinger R, Bogdanov-Berezovsky A, Silberstein E, Sagi A, Judkins K. (2005). Escharotomy using an enzymatic debridement agent for treating experimental burn-induced compartment syndrome in an animal model. *J Trauma.* 58: 1259-1264.

- Lamppa JW, Griswold KE. (2013). Alginate lyase exhibits catalysis-independent biofilm dispersion and antibiotic synergy. *Antimicrob Agents Chemother.* 57: 137-145.

- Leissring MA. (2008). The AβCs of Aβ-cleaving proteases. *J Biol Chem.* 283: 29645-29649.

- Masella RS, Meister M. (2006). Current concepts in the biology of orthodontic tooth movement. *Am J Orthod Dentofacial Orthop.* 126: 458-468.

- Pillai K, Akhter J, Chua TC, Morris DL. (2013). Anticancer property of bromelain with therapeutic potential in malignant peritoneal mesothelioma. *Cancer Invest.* 31: 241-250.

- Schreibelt G, Horssen JV, Rossum SV, Dijkstra CD, Drukarch B, de Vries HE. (2007). Therapeutic potential and biological role of endogenous antioxidant enzymes in multiple sclerosis pathology. *Brain Res Rev.* 56: 322-330.

- Shenoy AK, Ramesh KV, Chowta MN, Adhikari PM, Rathnakar UP. (2014). Effects of botropase on clotting factors in healthy human volunteers. *Perspect Clin Res,* 5: 71-74.

- Sporek M, Dumnicka P, Gala-Bladzinska A, Ceranowicz P, Warzecha Z, Dembinski A, Stepien E, Walocha J, Drozdz R, Kuzniewski M, Kusnierz-Cabala B. (2016). Angiopoietin-2 is an early indicator of acute pancreatic-renal syndrome in patients with acute pancreatitis. *Mediators Inflamm.* 2016: 5780903.

- Strohl WR. (2009). Therapeutic monoclonal antibodies: past, present and future. In: Therapeutic Monoclonal Antibodies from bench to clinic (Eds. Zhiqiang An) New York, USA: John Wiley & Sons, pp 4-50.

- Wei JM, Zhu MW, Zhang ZT, Jia ZG, He XD, Wan YL, Wang S, Xiu DR, Tang Y, Li J, Xu JY, Heng QS. (2010). A multicenter, phase III trial of hemocoagulase Agkistrodon: hemostasis, coagulation, and safety in patients undergoing abdominal surgery. *Chin Med J.* 123: 589-593.

- Xiang Y, Zeng H, Liu X, Zhou H, Luo L, Duan C, Luo X, Yan H. (2013). Thymidine kinase 1 as a diagnostic tumor marker is of moderate value in cancer patients: A meta-analysis. *Biomedical reports.* 1: 629-637.

GLOSSARY

Active Site: The region of an enzyme molecule that contains the substrate binding site for converting the substrate into product.

Albumin: A group of globular proteins, soluble in water but form insoluble coagulants when heated.

Allosteric Inhibitor: It binds to the enzyme at a site other than the substrate-binding site.

Amino Acids: The 20 basic building blocks of proteins, consisting of the basic formula R-CH (NH_2) -COOH, where "R" is the side chain which defines the amino acid.

Amylase: Enzyme that degrades starch, glycogen and other polysaccharides.

Angiogenesis: The process of formation of new blood vessels from pre-existing blood vessels.

Anion: A negatively charged ion, which moves towards anode.

Antibiotics: Natural substances secreted by microbes or their respective semi synthetic chemical forms that destroy or inhibit the growth of microorganisms, particularly pathogenic bacteria.

Antibody: Also known as an immunoglobulin (Ig), is a large, Y-shaped protein produced mainly by plasma cells that is used by the immune system to neutralize pathogens such as pathogenic bacteria and viruses.

Antigen: A toxin or other foreign substance which induces an immune response in the body, especially the production of antibodies.

Apoptosis: Programmed or organised death of cells which could be physiological or pathological.

Autoimmune Disorder: This occurs when the body's immune system attacks and destroys healthy body tissue by mistake. There are more than 80 types of autoimmune disorders.

Bacteria: Bacteria are single celled prokaryotic microorganisms, which live in a wide range of environmental conditions. Size ranges from 0.2 to 10 µm and they exhibit diverse shapes viz., rods, cocci etc.

Bromelain: A proteolytic enzyme, found in pineapples, used to treat inflammation

C Reactive Protein: C-reactive protein (CRP) is an annular, pentameric protein found in blood plasma, and its level rises in response to inflammation. It is an acute-phase protein of hepatic origin that increases following interleukin-6 secretion by macrophages and T cells.

Cancer: Cancer, also called malignancy, is an abnormal growth of cells. There is uncontrollable cell division leading to destruction of body tissue. The cancerous cells have potential to invade or spread to other parts of the body.

Carcino Embryonic Antigen (CEA): Carcinoembryonic antigen (CEA) is an oncofetal glyocoprotein, involved in cell adhesion. The blood level of this protein disappears or becomes very low after birth. In adults, an abnormal level of CEA may be a sign of cancer.

Carcinoma: It is used to describe cancer derived from epithelial cells that line various tissues throughout the body, eg., carcinoma of the breast, colon, liver, lung, pancreas

Cation: A positively charged ion, which moves towards cathode

Cirrhosis: Cirrhosis is a late stage of scarring (fibrosis) of the liver caused by many forms of liver diseases and conditions, such as hepatitis and chronic alcoholism.

Coagulation: Coagulation is the process by which blood changes from a liquid to a gel, forming a blood clot.

Coenzyme: An organic molecule that associates with enzymes and influences their activity.

Cofactor: A non-protein component essential for the normal catalytic activity of an enzyme. It may be an organic molecule or inorganic ion.

Collagen: Most abundant fibrous protein present in connective tissues viz., tendon, ligament, skin etc. It consists of three polypeptide chains and hence, it has a triple helical structure.

Collagenase: Enzyme that hydrolyses collagen

Competitive Inhibition: A type of enzyme inhibition, where inhibitor binds to the active site on enzyme, preventing the binding of substrate with enzyme.

Complement Proteins: The Complement system is an enzymatic mechanism in the blood containing nine proteins (termed C1-C9) which are produced in macrophages, liver cells and epithelial cells in the gastrointestinal mucosa. They then circulate in the bloodstream, until activated by antibodies causing a cascade of enzymatic reactions.

Covalent Bond: Relatively strong molecular bond in which the electronic configuration of the constituent atoms is fulfilled by sharing electrons.

Cryoglobulins: Cryoglobulins are single or mixed immunoglobulins that undergo reversible precipitation at low temperatures.

Dalton: A unit of mass equivalent to the mass of a hydrogen atom (1.66×10^{-24} g)

Denaturation: The disruption of the native structure of nucleic acid or protein molecule by heat, chemical treatment or changing pH conditions.

Diabetes Mellitus: A carbohydrate metabolism disorder which is characterized by impaired ability of the body to produce or respond to insulin, resulting in an increased blood glucose level.

Diagnosis: The process of identifying a disease, condition, or injury from its signs and symptoms.

Diagnostics: A procedure or equipment used to identify a disease or medical condition.

DNA (Deoxyribonucleic Acid):   A polydeoxyribonucleotide in which the sugar is deoxyribose; the main repository of genetic information in all cells and most viruses; usually double stranded.

Elastin: Highly extensible and chemically resistant yellow, fibrous protein, occurring in certain vertebrate connective tissue such as a fine network in the reticular dermis (corium) layer and around certain arteries, neck ligaments etc

ELISA: Enzyme linked immunosorbent assay - An immunoassay designed for detecting and quantifying substances such as peptides, proteins, antibodies and hormones. The detection strategy is mainly based on the highly specific antigen – antibody interaction.

Endonuclease: Cleaves bonds within a nucleic acid chain; they may be specific for RNA (RNAses) or for single-stranded or double-stranded DNA (DNAses). A restriction enzyme is a type of endonuclease.

Enzyme: A class of proteins that act as catalysts in biochemical reactions.

Escherichia Coli (E. coli):   A Gram negative bacterium commonly found in the vertebrate intestine; most frequently used in the study of biochemistry, molecular biology and genetics.

Factor VIII: A blood-clotting protein, also known as anti-hemophilic factor.

Ferritin: Ferritin is a universal intracellular protein that stores iron and releases it in a controlled fashion. It acts as a buffer against iron deficiency and iron overload.

Fibroblast: A prominent connective tissue cell, spindle shaped, which secretes collagen.

Fibrosis: It is the overgrowth, hardening, and/or scarring of various tissues and is attributed to excess deposition of extracellular matrix components including collagen.

Follicle Stimulating Hormone (FSH):   FSH is a gonadotropin, a glycoprotein polypeptide hormone. FSH is synthesized and secreted by the gonadotropic cells of the anterior pituitary gland and it regulates the development, growth, pubertal maturation and reproductive processes of the body.

Gastrin: Gastrin is a peptide hormone that stimulates secretion of gastric acid (HCl) by the parietal cells of the stomach and it aids in gastric motility.

Gene Therapy: It is a technique which includes insertion, deletion or alteration of genes for correcting defective genes responsible for disease development.

Gene: A hereditary unit consisting of a sequence of DNA that occupies a specific location on a chromosome and determines a particular characteristic in an organism; a segment of DNA coding for a specific protein.

Genetic Disorder: An inherited condition or problem caused by abnormalities in the genome.

Globulins: A group of globular proteins generally insoluble in water and are present in blood, eggs, milk, and as a reserve protein in seeds.

Glycoprotein: A glycosylated protein.

Growth Hormone: Growth hormone, also known as somatotropin, is a peptide hormone that stimulates growth, cell reproduction and cell regeneration in humans and other animals.

Hemoglobin: Hemoglobin (Hb) is a protein found in the red blood cells that carries oxygen and gives blood its red colour.

Hepatitis: Inflammation of the liver, mainly caused due to viral infection

Holoenzyme: The complete enzyme including all subunits often used with reference to RNA and DNA polymerases.

Hormone: Hormones are chemical messengers produced by endocrine glands and they control most of our body's major systems.

Human Chorionic Gonadotropin (hCG):   hCG is a hormone produced by the placenta after implantation. The presence of hCG is detected in some pregnancy tests.

Immobilized Enzyme: An enzyme fixed by physical or chemical means to a solid support e.g. a bead or gel. It could be easily separated and recovered from the products and used again.

Immune System: *Immune system* defends our body against foreign substances, cells, and tissues by producing the immune response. Thymus, bone marrow and lymph nodes are the major organs of immune system.

Immunity: The ability of an organism to resist a particular infection or toxin by the action of antibodies.

Immunoglobulins (Ig): Ig is a protein produced by plasma cells and lymphocytes. It gets attached to foreign substances, such as bacteria, and assists in destroying them. There are five immunoglobulin classes found in serum: IgG, IgM, IgA, IgE and IgD.

Infection: Infection is the invasion of an organism's body tissues by disease-causing agents, their multiplication, and the reaction of host tissues to the infectious agents and the toxins they produce.

Inflammation: A local immune response to cellular injury that is marked by capillary dilatation, leukocytic infiltration, redness, heat, pain etc

Insulin: A hormone that controls blood sugar. It is produced in the islets of Langerhans (pancreatic islets), which are small isolated clumps of special cells in the pancreas.

Isoelectric Point or pH: The pH at which a protein has no net charge.

Isomerase: An enzyme that catalyzes an intramolecular rearrangement.

Isozymes: Multiple forms of an enzyme that differ from one another in one or more of their properties.

Keratin: Fibrous scleroprotein present in hair, feathers, hooves and horns.

Kilobase (kb): A unit used at the molecule level for measuring distances among nucleic acids, chromosomes or genes equal to 1000 bases (equivalent to 1000 nucleotides or base pairs).

Ligand: An ion or molecule that donates a pair of electrons to a metal atom or ion to form a coordination complex.

Lipase: Fat-splitting enzyme that catalyzes the hydrolysis of fats or lipids.

Lipoprotein: Lipoproteins are lipid-protein complexes that allow all lipids derived from food or synthesized in specific organs to be transported throughout the body by the circulatory system.

Luteinizing Hormone (LH): LH is produced by gonadotropic cells in the anterior pituitary gland and it is one of the main hormones that control the reproductive system.

Lyase: Enzymes that catalyse the non-hydrolytic addition or removal of groups which are free during reaction.

Lymphocytes: A lymphocyte is one of the subtypes of white blood cells in the vertebrate's immune system. They are the main type of immune cells found in the lymph. The three major types of lymphocytes are T cells, B cells and natural killer (NK) cells.

Lysozyme: An enzyme occuring naturally in egg white, human tears, saliva, and other body fluids, capable of destroying the cell walls of certain bacteria and thereby acts as an antiseptic. It is also used for cancer chemotherapy.

Macrophages: White blood cells, which are an integral part of the immune system. They engulf the foreign particles or infectious organisms by the process of phagocytosis.

Metalloprotease: It is a group of proteases whose functional group in their active site involves a metal.

Michaelis Constant (K_m):   The substrate concentration at which an enzyme-catalyzed reaction proceeds at one-half of the maximum velocity.

Monoclonal Antibodies (Mabs):   Antibodies that are produced by identical clone of cells. Mabs are highly specific in binding to antigen epitopes.

Myocardial Infarction: Commonly known as heart attack. It is the damage caused to the heart muscle (myocardium) due to insufficient blood supply.

Native Gel: An electrophoresis gel run under conditions which do not denature proteins (i.e., in the absence of SDS, urea, 2-mercaptoethanol, etc.).

Necrosis: Dell death occuring due to infectious agents, hypoxia, extreme environmental conditions etc.

Noncompetitive Inhibitor: An inhibitor of enzyme activity whose effect is not reversed by increasing the concentration of substrate molecule.

Nucleus: The large body embedded in the cytoplasm of all plant and animal cells containing the genetic material. It functions as the control centre of the cell.

Osteoblasts: Bone forming cells

Osteoclasts: Type of large multinuclear bone cells closely associated with bone resorption.

Papain: A cysteine protease present in papaya and is also known as papaya proteinase I.

Parathyroid Hormone (PTH): PTH also called parathormone or parathyrin, is a hormone secreted by the parathyroid glands. It is important in bone remodeling, which is an ongoing process in which bone tissue is alternately resorbed and rebuilt over time.

Peptide: A chain formed by two or more amino acids linked through peptide bonds.

Plasma: Plasma is the noncellular fraction of blood which transports red cells, white cells and platelets. It is an aqueous solution of salts, proteins, lipids, sugars and various other substances serving a myriad of physiological functions. It is rich in protein and contains clotting factors.

Platelets: Platelets, also called thrombocytes, are a component of blood. Its function is to react to bleeding from blood vessel injury by clumping, thereby initiating a blood clot.

Polar Group: A hydrophilic (water-loving) group.

Polyacrylamide Gel Electrophoresis (PAGE): Technique used to separate proteins and smaller DNA fragments and oligonucleotides by electrophoresis.

Primary Structure: In proteins, it refers to the amino acid sequence.

Prognosis: Forecasting the probable course and outcome of a disease, especially the chances of recovery.

Prolactin: Prolactin, also known as luteotropic hormone or luteotropin, is a protein secreted by the pituitary gland which enables lactation (production of milk) in mammals, usually females.

Prostate Specific Antigen: Prostate-specific antigen (PSA), also known as gamma-seminoprotein or kallikrein-3 (KLK3), is a glycoprotein encoded in humans by the KLK3 gene. It is secreted by the epithelial cells of prostate gland.

Prosthetic Group: A tightly bound non-peptide inorganic or organic component of a protein. Prosthetic groups may be lipids, carbohydrates, metal ions, phosphate groups etc.

Protease: An enzyme that performs proteolysis; also termed peptidase or proteinase

RBC: Red blood cells or erythrocytes are a type of blood cells which are highly specialized for their primary function of transporting oxygen from lungs to all body tissues.

Recombinant Protein: Proteins encoded by recombinant DNA that has been cloned in a foreign expression system to support the expression of the exogenous gene.

Renin: Renin, also known as an angiotensinogenase, is an aspartic protease secreted by the kidneys that participates in the body's renin–angiotensin–aldosterone system. It regulates the body's mean arterial blood pressure.

Rheumatoid Factor: Rheumatoid factors (RF) are autoantibodies produced by the immune system that can attack healthy tissues in the body. High levels of rheumatoid factor in the blood are most often associated with autoimmune diseases, such as rheumatoid arthritis and Sjogren's syndrome.

Ribozyme: A catalytically active RNA.

RNA (ribonucleic acid): A polynucleotide in which the sugar is ribose.

RNA Polymerase: An enzyme that catalyzes the synthesis of RNA from ribonucleotide triphosphates, using DNA as a template.

SDS PAGE: This electrophoretic method is based on the separation of proteins according to size; SDS treatment assigns uniform negative charge to proteins.

Secondary Structure: In a protein or a nucleic acid, any repetitive folded pattern that results from the interaction of the corresponding polymeric chains. In proteins, the most common are α helix and β pleated structures.

Serum: It is the clear yellowish coloured liquid that gets separated from clotted blood. It is protein rich, but devoid of clotting factors and blood cells.

Substrate: A molecule that is acted upon, and chemically changed, by an enzyme.

Subunit: Individual polypeptide chains in a protein.

Tertiary Structure: With reference to protein or nucleic acid, it is the final folded form of the polymer chain.

Therapeutic: Things which help to treat or cure a disease

Thyroid Stimulating Hormone (TSH):   Also known as thyrotropin, thyrotropic hormone. TSH is a pituitary hormone that stimulates the thyroid gland to produce thyroxine (T4) and triiodothyronine (T3) which stimulate the metabolism of almost every tissue in the body.

TNF - α: Tumor necrosis factor α, is a multifunctional cytokine involved in a wide spectrum of biological processes including cell proliferation, differentiation, apoptosis, lipid metabolism and coagulation. It is mainly produced by the macrophages and is involved in the fight against tumors.

Total Protein: Serum total protein assay measures the total amount of protein present in the blood.

Transamination: A biochemical reaction in which an amine group is transferred from an amino acid to keto acid to form a new amino acid and keto acid. The coenzyme required for this reaction is pyridoxal phosphate.

Transferase: An enzyme that catalyzes the transfer of a functional group from one molecule to another.

Transport Protein: A protein whose primary function is to transport a substance from one part of the cell to another (organelles), from one cell to another, or from one tissue to another.

Troponin: Troponin or the troponin complex is a complex of three regulatory proteins (troponin C, troponin I and troponin T) that is integral to muscle contraction in skeletal muscle and cardiac muscle, but not smooth muscle.

Trypsin: A proteolytic enzyme that cleaves peptide chains next to the basic amino acids arginine and lysine.

USFDA: Food and Drug Administration is a federal agency of United States Department of Health and Human Services. It is responsible for protecting and promoting public health through the control and supervision of food safety, pharmaceutical drugs, vaccines, medical devices etc.

Vaccine: Vaccine is a biological preparation that provides immunity to a particular disease. Vaccine typically contains an agent that resembles a disease-causing microorganism, and is often made from weakened or killed forms of the microbe, its toxins or one of its surface proteins.

Virus: A complex of nucleic-acid and protein, that can infect and replicate inside a specific host cell to make more virus particles. Size of virus ranges from 20 to 400 nm and exhibit different shapes viz., helical, polyhedral etc

Vitamin: Organic compounds required by living organisms in relatively small amounts to maintain normal health.

WBC: White blood cells (also called leukocytes or leucocytes) are the cells of the immune system that are involved in protecting the body against both infectious disease and foreign invaders.

Western Blot: A technique used to detect the presence of a specific antigen (protein) on a nitrocellulose membrane with the help of specific antibodies.

Zwitterion: A dipolar ion with spatially-separated positive and negative charges. e.g. most amino acids are zwitterions, having a positive charge on the α-amino group and a negative charge on the α-carboxyl group but no net charge on the overall molecule.

Zymogen: An inactive precursor of an enzyme. e.g. trypsin exists in the inactive form trypsinogen before it is converted to its active form, trypsin.

Index

Made in the USA
Monee, IL
07 July 2026

56551751R00129